Energy Security and Natural Gas Markets in Europe

T0227466

Moving beyond most conventional thinking about energy security in Europe, which revolves around stability of supplies and the reliability of suppliers, this book presents the history of European policy making regarding energy resources, including recent controversies about shale gas and fracking. Using the United States as a benchmark, the author tests the hypothesis that European Union energy security is at risk primarily because of a lack of market integration and cooperation between member states. This lack of integration still prohibits natural gas to flow freely throughout the continent, which makes parts of Europe vulnerable in case of supply disruptions.

The book demonstrates that the European Union gas market has been developing at different speeds, leaving the northwest of the continent reasonably well integrated, with sufficient trade and liquidity and different supplies, whereas other parts are less developed. In these parts of Europe there is a structural lack of investments in infrastructure, interconnectors, reverse flow options and storage facilities. Thus, even though substantial progress has been made in parts of the European Union, single-source dependency often prevails, leaving the relevant member states vulnerable to market power abuse. Detailed comparisons are made of the situations in the Netherlands and Poland, and of energy policy in the USA. The book dismantles some of the existing assumptions about the concept of energy security, and touches upon the level of rhetoric that features in most energy security and policy debates in Europe.

Tim Boersma is a Fellow and Acting Director of the Energy Security and Climate Initiative, part of the Foreign Policy Program at the Brookings Institution, Washington DC, USA. He has a PhD in international relations from the University of Groningen, the Netherlands, and has previously been a Research Fellow, Transatlantic Academy, Washington, DC, and Visiting Scholar at the School of Advanced International Studies, Johns Hopkins University.

Routledge Studies in Energy Policy

Our Energy Future
Socioeconomic implications and policy options for rural America
Edited by Done E. Albrecht

Energy Security and Natural Gas Markets in Europe
Lessons from the EU and the USA
Tim Boersma

International Energy Policy
The emerging contours
Edited by Lakshman Guruswamy

For further details please visit the series page on the Routledge website:
www.routledge.com/books/series/RSIEP/

Energy Security and Natural Gas Markets in Europe

Lessons from the EU and the United States

Tim Boersma

Routledge
Taylor & Francis Group

LONDON AND NEW YORK

earthscan
from Routledge

First published 2015 by Routledge

2 Park Square, Milton Park, Abingdon, Oxfordshire OX14 4RN
711 Third Avenue, New York, NY 10017

Routledge is an imprint of the Taylor & Francis Group, an informa business

First issued in paperback 2017

British Library Cataloguing-in-Publication Data
A catalogue record for this book is available from the British Library

Library of Congress Cataloging in Publication Data
Boersma, Tim.
Energy security and natural gas markets in Europe : lessons from the EU and the United States / Tim Boersma.
pages cm. — (Routledge studies in energy policy)
Includes bibliographical references and index.
1. Energy security—European Union countries. 2. Energy security—United States. 3. Energy policy—European Union countries. 4. Energy policy—United States. I. Title.
HD9502.E852B64 2015
333.79094—dc23
2015012366

ISBN: 978-1-138-79512-9 (hbk)
ISBN: 978-1-138-57491-5 (pbk)

Typeset in Sabon
by FiSH Books Ltd, Enfield

To Ieke, Susana, Nynke, Jan, and Sjoerd

Contents

Acknowledgements

This book is an expanded version of my dissertation manuscript, which I defended in September 2013 at the University of Groningen. I am grateful to Routledge for giving me the opportunity to reach a wider academic audience, and to the Brookings Institution for providing me with an environment in which I could spend a significant amount of time updating and adjusting the manuscript. And I am grateful to so many friends and acquaintances in the industry and policy community who have stimulated and challenged me throughout the years.

I want to thank my former colleagues at Brabers corporate counsel in The Hague for giving me the opportunity to conduct research in their corporate setting. I am also grateful to my former colleagues at the German Marshall Fund of the United States. In addition, I am also filled with gratitude for the inspiring (and ongoing) conversations that I have had with my colleagues at the Brookings Institution. Specifically I would like to thank Charles Ebinger, for giving me the opportunity to be part of the Brookings family, an environment in which I learn on a daily basis. I also thank the (former) ESCI staff for fruitful conversations and comments, and invaluable support. Jennifer Potvin, Heather Greenley, Colleen Lowry, Cameron Khodabakhsh, thank you. Special thanks to Christine Johnson for spending so many hours getting the manuscript in its right form. Thanks to Geert Greving for our enjoyable exchanges that have been so informative. I want to express my appreciation for my colleagues outside ESCI, for interesting conversations, debates, and differences of opinion. Fiona Hill, Cliff Gaddy, Hannah Thoburn, Andrew Moffatt, Jeremy Shapiro, Thomas Wright, Tanvi Madan, Kemal Kirisci, Natan Sachs, Erica Downs, Kevin Foley, Jonathan Pollack, Steven Pifer, Bruce Jones, thank you. I am grateful to Martin Indyk and Ted Piccone, both former Vice Presidents of Foreign Policy at Brookings, for their support.

For reviews of earlier versions and the current manuscript, I am very grateful to Jaap de Wilde, Catrinus Jepma, Leigh Hancher, Coby van der Linde, Herman Hoen, Stacy D VanDeveer, Charles Ebinger, Michael O'Hanlon, and two anonymous reviewers.

Last and certainly not least, I want to thank my family, Ieke, Jan, Sjoerd, and Nynke, my dear friends, and my lovely wife, Susana. You all rock.

Summary for policy makers

Contrary to the audience intended for this book, this summary is designed for policy makers in European capital cities, including Brussels, and, to a lesser extent, Washington, DC. At the request of the publisher, the book was written for an academic audience. It may seem surprising, in the midst of the fierce debate among pundits, policy makers and energy commentators about diversification away from Russian gas supplies, that the lessons for policy makers, derived from a review of the energy security literature, extensive fieldwork, and decades of market and institutional development in the European Union, are plain and simple.

First, integrating EU energy markets and enhancing cooperation between member states is the most efficient tool to enhance energy security. The available evidence is overwhelming. In the larger part of the EU, member states have in recent decades increasingly integrated their markets, and by doing so have gained access to various sources of supply. It is in these countries that dependence on a single source is no longer an issue. In particular, in light of the Ukraine crisis, which seems to have brought geopolitics right back on top of the agenda, it is crucial to keep in mind that zero-sum thinking, tempting as it may be, is not the way forward.

It is, however, also important to note that the standard EU mantra of liberalizing markets and implementing legislation can no longer be considered a panacea. It is about time that decision makers in northwestern Europe, including Brussels, acknowledge that in certain parts of the EU additional incentives to invest have to be adopted, in order to enhance energy security. In recent years, several prominent examples of this have come forth. Poland, which has, rightly or not, been one of the most vocal member states when it comes to the dangers of dependence on natural gas from Russia, has in just a couple of years' time and with the support of EU funding, developed additional interconnection facilities with neighboring countries, most prominently Germany. It has also initiated the construction of a liquefied natural gas (LNG) terminal on its northern shores, which is expected to come on-stream later in 2015. The economics of some of these projects, and certainly the LNG terminal, can be questioned. Nevertheless, once the LNG terminal comes on-stream, the Polish TSO estimates that around 90 percent of Polish natural gas demand can be covered by sources other than Russian gas. That of course does not mean that Poland will no longer consume Russian gas. To the contrary, buyers in Poland, like anywhere else, will buy the cheapest available feedstock, which in this case is Russian gas. The point, however, is that there is access to alternatives, and that is what matters. Similarly, the Czech Republic has integrated its market successfully with the German market. Here too, single source dependency is solved. Finally, in Lithuania the construction of an LNG terminal alone incentivized the renegotiation of existing long-term oil-indexed contracts with Gazprom, and Lithuanian prices came down substantially. Here too the economics of the project were not decisive, and EU subsidies and possible amendments to existing domestic legislation were necessary to make the project viable. These projects, however, should be seen as (pricey, arguably) insurance policies to prevent possible market power abuse by a dominant supplier, in this case Russian Gazprom.

For too long policy makers in northwestern Europe have argued that the market would take care of energy security, if only existing legislation were implemented. I believe that this is not always the case, and acknowledge that a little push is in certain cases required. As illustrated throughout this book, there are a number of other countries in central and eastern Europe (e.g. Bulgaria and Romania) where a similar line of reasoning applies. This is also an important lesson when it comes to thinking of new incentives to attract new investments in energy infrastructure, which in parts of Europe are direly needed. The tug-of-war between European institutions and member states about the decision-making power when it comes to energy policy will not end anytime soon. All the existing infrastructural bottlenecks have been documented extensively and debated for years, and so it is time that EU and member state decision makers put their money where their mouth is, and direct the available financial means to the places where it is most urgently needed. This is in only a handful of countries, namely they are Bulgaria, Hungary, Romania, and to a lesser extent Greece and Croatia. It is worth noting that in the wider context of the EU Energy Community the barriers are more widespread and persistent, but this book only focuses on the EU internal market.

It requires political leadership in the other member states to acknowledge that these few member states is where collective financial means should flow, and not to domestic pet projects that politicians generally tend to advocate. In addition, it is also common knowledge that the existing financial means that European institutions have available are not sufficient to address the existing bottlenecks. The only structural mandate that the European Commission has to co-invest in energy infrastructure is under the Connecting Europe Facility. Available financial means comprise a modest €5.85 billion for the period until 2020, which is only 3 percent of the estimated required investments (in both natural gas and electricity infrastructure). At this point one can only hope that the plans for an Energy

Union, which are debated as this book is in press, also contain new proposals to attract investments in energy infrastructure, because further expanding the financial bandwidth of the European Commission is unfeasible, and outright public financing of infrastructure is a questionable proposition to begin with. Possibly the European Investment Bank can play a more prominent role here, or more generous rates of return for specific projects can be applied, in order to attract pension funds or institutional investors (a model that has been applied in individual member states, such as Italy).

Nevertheless, when considering the progress that has been made over the last half decade in terms of market integration, it cannot be disputed that this is the way forward to solve EU energy security concerns. Resilience today is significantly higher than it was only several years ago, and the member states did not stop buying particular kinds of natural gas. That puts all the rhetoric about Russian gas in perspective, and this is important to keep in mind when designing future policy.

The second major lesson that follows directly from this point is that price matters. What some observers and politicians tend to forget is that the EU has slowly but gradually moved towards becoming a liberalized gas market. In this context it is private or privatized actors who purchase and trade commodities. Though these actors have to operate in a political climate that may well affect their decisions, political preference does not become part of the commercial lexicon. This has a number of important consequences. First, this has to be taken into account when analyzing the (sometimes harsh) day-to-day rhetoric about natural gas from Russia. Also in eastern Europe, the case of Poland demonstrates that at the end of the day the cheapest supplies of natural gas are preferred over other sources. The aforementioned LNG project was only constructed with the help of subsidies and regulatory policies. It remains to be seen to what extent the plant will be used to bring in natural gas. Indirectly, the Polish tax payer will pay a premium for natural gas that is imported in the form of LNG. Evidently energy security comes at a cost, yet these costs vary throughout the EU. Second, this has to be taken into account when examining the efforts of EU policy makers to strengthen ties with alternative suppliers, for instance from the Caspian region or the eastern Mediterranean. Attractive as it may sound to purchase natural gas from exotic places, it is important to keep in mind that unless European buyers collectively decide that they are willing to pay a premium for their natural gas, the cheapest commodity will always be preferred. In the case of Europe, it seems that for the foreseeable future that means natural gas that is domestically produced, and then imports from Russia, Norway, and Algeria will be most in demand. Whatever additional supplies are required, can be imported in the form of LNG. To give an indication of the premium that would have to be paid, estimates from Bernstein Research suggest that approximately 57 billion cubic meters (bcm) of natural gas demand can be saved, and thus Russian imports reduced, at the cost of US\$33 billion per year. Measures would include drawing down gas inventories, outbidding Asia on LNG and switching to oil as a feedstock for electricity generation. When it comes to outbidding Asian markets for LNG supplies, estimates suggest that around 18 bcm of Russian imports per year could be replaced, at an annual cost of US\$5 billion. This assumes that Asian contracted LNG can be diverted at a cost of US\$17 per million British thermal units (MMBtu). Most studies suggest that in the coming years the amount of imports of LNG into the EU will increase substantially. Regardless, it is important to keep in mind that because of the low marginal costs of Russian gas Gazprom likely holds the marketing power in the foreseeable future. Thus it is important to note that additional LNG will mostly replace dwindling domestic production. It is also important to reiterate that for the places where alternative supplies may be most direly needed, the existing infrastructure is not in place to actually get the LNG there. This again confirms the urgent need to complete the integration of the internal gas market, so that all member states have access to several sources of supply. This can be done by targeted investments in infrastructural bottlenecks, implementation of existing legislation, and where necessary better streamlining of between national regulatory approaches.

As an import dependent continent, the primary concern of EU decision makers has to be how to design a gas market that is attractive for outside suppliers. In the context of globally rising demand for natural gas, the EU cannot afford to haggle over the origins of its natural gas. A European Commissioner recently stated that imports from countries with an autocratic rule cannot be accepted anymore (referring to imports from Russia). Fine, yet in the same talk it is announced that ties with suppliers in countries like Algeria, Azerbaijan, Turkmenistan, and others have to be strengthened. This does not make sense, and unless EU buyers decide to pay a significant premium for natural gas from preferred sellers, then this rhetoric should be toned down. Modelling efforts carried out on behalf of the EU in 2014 demonstrated that the share of natural gas in the EU (roughly 23% of primary energy consumption) can be reduced further if desired. Options of fuel switching in electricity generation are somewhat limited: given the low cost of coal most existing coal-fired electricity generation is utilized at this point. There is ample room to increase the share of renewable energy. Further energy efficiency measures can drive down demand for natural gas. The question is whether this is where the focus of the EU should be. The EU has sufficient barriers to overcome domestically, in terms of ironing out regulatory flaws, and making natural gas an attractive feedstock to begin with. It could further develop technology and regulatory regimes to tap into its unconventional natural gas potential. Though this falls outside the scope of this book, in the current constellation natural gas is in a tough spot in the EU, with low carbon prices, low coal prices, and subsidies for renewables making natural gas a less attractive feedstock. If EU leaders want natural gas to play a prominent role in its fuel mix, and given the ambitious renewables agenda there are good reasons to assume that, then the focus should be on completing and fine-tuning the internal market, rather than quarrelling about one external supplier. That is a political debate that is increasingly distant from market realities. As described, the current trajectory to address concerns of market abuse by certain suppliers (i.e. namely by further European cooperation and integration) has proven to be the right one and should be continued.

1 Europe's focus on available natural gas supplies, and why the focus is too narrow

For decades natural gas was not an issue of major concern. Apart from squabbles between the first Reagan Administration and European leaders about increasing dependence of Europe on (then) Soviet natural gas, energy in general had a remarkably low profile in the period between the oil crises and the mid-2000s (see McGowan 2011: 491). This changed when Ukraine and Russia could not solve a price dispute, and the former decided in January 2006 to seize supplies of natural gas to the EU.

Much has been written about who is to blame for this supply disruption and the one that followed in 2009, how real or perceived the risks are and subsequently what response options the EU and/or the individual member states has/have (e.g. Smith Stegen 2011; Schmidt-Felzmann 2011; Le Coq and Paltseva 2012). This debate has once again been revitalized in light of the Ukraine crisis that started in early 2014, even though to date natural gas continues to flow to European buyers.[1] It is worth noting that by now the EC openly acknowledges that the problem of supply disruptions and an acute lack of natural gas as experienced in the past could have been solved if the internal energy system would have been functioning better. This would notably be the case with more interconnection capacity, reverse flow options and storages facilities (European Commission 2010). Therefore, this book is not about Russia. It is also not about external suppliers more broadly. This book is about the cooperation between European member states on energy policy. It is about governance within the EU. It has that specific focus, because the available empirical evidence, as discussed in more detail throughout this book, overwhelmingly suggests that cooperation between European member states solves the issue of dominant external suppliers. Yet, despite that reality, this book also shows that member states remarkably struggle to cooperate on energy issues. This book delves into these governance questions. As an indication that this issue is relevant, the Eurasia Group labelled the politics of Europe as the top political risk for 2015.[2]

This analysis is based on the premise that gas security is safeguarded by access to diverse supplies, and the ability to have those flow freely throughout the continent. In order to facilitate the latter, sufficient gas infrastructure is paramount. An adequate organization of gas distribution systems in Europe has been on the agenda for quite some time (Yergin 1988). It is also part of ongoing discussions about a concept that the European Commission led by President Juncker in October 2014 has labelled an "Energy Union."[3] Hence the question is validated why apparently up to date the internal European gas system is not always functioning properly.

This book assesses whether certain preconditions for efficient market functioning are in place. This is in contrast with more accepted econometric approaches of ex post price convergence analysis (e.g. Neumann and Siliverstovs 2005; Renou-Maissant 2012). These studies generally assess market integration by making an ex post analysis of prices in different markets. Over time converging prices are seen as a sign of well integrated markets. Several commentators have concluded that parts of the EU gas system are already reasonably well integrated (Harmsen and Jepma 2011; Renou-Maissant 2012; Heather 2012). Yet in other parts of the EU this is arguably not the case, for a variety of reasons that are discussed in detail later. The institutional approach taken here aims to answer the basic notion whether natural gas can, under current conditions, flow throughout the EU. This may be required – if not dictated by price – because of the risk of a supply disruption. Instead of an *ex post* analysis, this book aims to determine *ex ante* whether several components of the European gas system are developed to the extent that the market can function properly. In addition this approach aims to determine whether existing decision-making structures in the EU are adequate to allow for these components to develop properly.

Though predictions have to be taken with a grain of salt, with the increase of liquefied natural gas (LNG) transport and also the development of unconventional natural gas, it is generally assumed that natural gas is abundantly available in the foreseeable future (International Energy Agency 2012a, 2012b, 2013: 107). Söderbergh et al. (2009, 2010) warned for possible limitations to future natural gas production in the two largest suppliers of the EU (i.e. Russia and Norway), due to uncertainties regarding the required yet costly development of new gas fields in Arctic Russia and the Far East, and a depletion in the potential for increased exports from Norway. On the other hand unconventional gas extraction, which is the topic of Chapter 6 of this book, is expected to substantially increase global supplies and its exports from countries as diverse as the US, Australia, and Russia.[4] With the existing uncertainties it is difficult to assess exactly what amount of natural gas is going to be available on global markets. What is certain, however, is that the EU in this equation is going to be a net importer: the Netherlands and Denmark are the only net exporters on the continent and their production is in decline, while the United Kingdom and Romania are currently by and large self-sufficient (European Commission 2012). As is discussed in Chapter 6 it is uncertain when and even whether large amounts of unconventional natural gas will be produced in Europe. However, it is safe to assume that even when that would eventually happen, large amounts of natural gas have to be imported. The Joint Research Center estimated in September 2012 that even if the unconventional natural gas potential in the EU were to be developed, this would only be sufficient to halt European import dependence at around 60 percent in the future (Pearson *et al.* 2012).

The EU gas system consists of four facets. These are, in arbitrary sequence: markets, infrastructural companies, governmental institutions and regulatory authorities. The market place is where the producers, suppliers, traders and consumers operate. This is where natural gas is supplied to both small consumers (retail) and large energy-intensive industries and where traders operate at energy exchanges and increasingly trade short-term (spot market) and long-term (futures) products. Infrastructural companies are in general publicly oriented companies. The most important reason for this characteristic is their natural monopoly position as the administrator of gas infrastructure. Governmental institutions, both national and supranational, set the ground rules for the playing field in which market players operate. Finally regulatory authorities monitor market players' behavior, guard over fair competition and decide over tariff changes and various other parameters in terms of for instance costs for the usage of infrastructure or appropriate technical standards.

Whether the EU member states can all equally benefit from abundant natural gas supplies is less certain, primarily due to slow market integration. The risks arising from the status quo regarding the EU natural gas system are at the basis of this book. As this analysis shows, a part of these risks stems from a mismatch in policy making within the trajectory of liberalization of the gas market. While markets have been liberalized and operate primarily on a European level, other essential parts of the energy system (notably regulation and infrastructure) only occasionally do, though in parts of Europe the situation has improved substantially.

An examination of government policies aimed at these different facets of the energy system shows remarkable differences. Within energy markets a clear trend of up scaling to the European level can be identified under influence of a series of gas directives.[5] In comparison, infrastructural companies and regulatory authorities operate mainly in their national domains. This mismatch in levels of government intervention within the EU energy system is demonstrated and discussed in more detail in the third chapter of this book. The second chapter shows that this mismatch, with a heavy focus on available natural gas supplies, is reflected in the academic debate on energy security as well. This suggests that a more integrated analysis of energy security in the EU is useful, taking into consideration all elements of the European gas system as just described. Moreover, the focus on available natural gas supplies is remarkable, given the wide availability of the resource.

The most important legislative documents regarding the EU internal market for natural gas come from European political institutions, in other words the EC and in some cases the European Parliament (EP). Their implementation is a matter of the member states. Different interpretations of Directives and different pace of implementation have caused friction within the EU. An example of this is the case of so-called unbundling of integrated energy companies. Some member states, such as the United Kingdom and the Netherlands (Box 1.1), have implemented this legislation more energetically than others, such as France and Germany. The EC has never made a secret of her intentions, namely that ideally within the EU all integrated companies should ultimately be ownership unbundled.[6] Contrary to many market players, regulatory authorities and infrastructural companies continue to operate predominantly at the national level. It is therefore hypothesized throughout this analysis that existing decision-making structures within the EU energy system are not always optimal. This is tested in different case studies, which are discussed later. To limit the scope of this book the focus is on the EU gas system, though relevant parallels with the electricity system exist.

As Chapter 2 demonstrates, most of the academic debate about energy security revolves around the availability of sufficient supplies, reliability of suppliers and/or supply routes. However, energy security cannot be labeled as a question of sufficient supplies or resources alone, given the vast current proven reserves/(un)conventional potential. For the EU it is unmistakable that future gas supplies increasingly consist of pipeline flows (from in particular Russia, Norway and Algeria) and LNG imports. Domestic production is in decline. This book comprises the notion that future EU energy security is not only a matter of sufficient supplies or reliability of suppliers, but also a matter of the ability to transport the commodity throughout the EU to its destination.

A growing number of studies make reference to the US gas system, albeit mainly with reference to investments in gas infrastructure (Von Hirschhausen 2008) or for market structure and functioning more in general (Ascari 2011; Vazquez *et al*. 2012). There is a widely held view that this is the only integrated and well-functioning gas system in the world (De Vany and Walls 1994). There is much less agreement whether the EU can or should aim to develop itself in a similar direction as the US has done, as is discussed later in this book. Yet it is argued here that there are two important reasons to involve the US gas system in this analysis. The first has been mentioned briefly: globally the US gas system is perceived as being the only well-functioning gas system. Hence, despite the myriad of differences (which are discussed in the case studies) and acknowledging that the US system is not perfect either, there may be lessons to be learned for Europe. Moreover, some recent developments in the EU, most notably the increase of spot-market trade and the decline of long-term oil-indexed contracts, confirm that European policy makers are actively looking at the US model

Box 1.1 Ownership unbundling in the Netherlands

Taking a closer look at the Netherlands' energy history, the government has been rather active in implementing both Electricity and Gas Directives. Yet comparing the gas market and the electricity market, some differences remain and lines between public and private activities sometimes remain thin.

The national grid operator for electricity TenneT is a wholly owned public company focussing solely on grid activities for networks of 110 kV and up. When purchasing German Transpower of E.ON AG in 2010 it became the first grid operator for electricity that crossed its national border. TenneT is the major shareholder (holding 56.1 percent of the shares) in the Dutch-Belgian energy exchange APX-ENDEX. The aim is to offer platforms to help increase the liquidity of the market.

In the Dutch gas market – as in many gas markets – the situation is complicated. Until 2005 all Dutch activities in the gas market were concentrated under the umbrella of NV Nederlandse Gasunie, a public-private partnership between the Dutch state, Shell and ExxonMobil. With the implementation of the Second Gas Directive, this partnership was divided into a trading company called GasTerra and an infrastructural company called Gasunie. The latter is, however, not entirely publicly oriented, for its transportation duties are carried out by Gas Transport Services, or GTS, but Gasunie is also involved in commercial gas storage, for instance through Zuidwending in Veendam, and participates in an LNG terminal (i.e. the GATE terminal in Rotterdam). Finally Gasunie holds 20.1 percent of the shares in energy exchange APX-ENDEX and 9 percent of the shares in NordStream.

Often the Dutch Gasunie is referred to as the public Dutch infrastructural company and in fact it is. However, some of its activities remain commercial.

as a model for the future. Second, the US has a federal structure, and as such shows resemblance with the European model of governance. It is uncertain whether the EU will eventually develop into a federation, but this can be considered as one plausible scenario. Therefore, examining and where possible comparing existing decision-making structures in these two gas systems is relevant. In line with Makholm (2012) it is hypothesized that the European gas system is currently not functioning properly, because too many issues are addressed at a suboptimal level of policy making. This can be because institutions are not developed, or their current mandates inadequate.

This book consists of three main parts. The first part, namely Chapter 2, lays out the analytical framework. First, a state of the art overview of energy security literature is presented. Admittedly, a wide though incomplete selection of literature is by definition arbitrary to an extent. Yet the overview demonstrates that the bulk of academic contributions on energy security focus on diversification of supplies and (unreliable) suppliers. These fall in the category "markets" that has been described earlier, as part of the energy system. Fewer contributions, however, focus on infrastructure and regulation. Therefore, indirectly this theoretical overview justifies the partial focus of this research on those two facets of the European gas system. Chapter 2 then proceeds with an overview of (neo)functionalist theory and also touches upon new institutional economics literature to answer for both the *ex ante* analysis of institutions in the European gas system, as well as underline the importance (and to an extent functioning) of decision-making structures in the EU. Subsequently Chapter 2 concludes with an overview of multilevel governance, which is presented as a suitable framework for analysis more than a theory of European integration. The latter debate falls outside the scope of this book and is therefore only briefly touched upon. The section does answer for the choice of this framework of analysis, its limitations and also its suitability for application to the benchmark (i.e. the US gas system).

Part II of the book analyses existing energy policy in Europe. Chapter 3 comprises an overview of existing European energy policies. It sheds light on existing regulations and mandates for European institutions and individual member states. It is important to note that it focuses on the internal market, rather than external suppliers. Chapter 4 looks specifically at Regulation 994/2010, which was designed to improve security of supply in the EU. This chapter assesses what contribution to security of supply this specific regulation has made since it was agreed on by the member states.

Part III contains three case studies. One of the most urgent problems for parts of the gas system in the EU is generating sufficient investments in infrastructure (Pelletier and Wortmann 2009; European Commission 2011). Therefore, Chapter 5 examines the investment climate for gas infrastructure, and related decision-making structures. Many of the current uncertainties in global gas systems can be linked to a phenomenon that has widely been labeled as "the shale gas revolution" (Trembath *et al.* 2012; Boersma and Johnson 2012). In addition, some of the European member states indeed may have substantial recoverable reserves of so-called unconventional natural gas in their soils. Therefore, shale gas extraction is the subject of Chapter 6. Some scholars have been rather optimistic about the ability of European institutions to mobilize member states to accept a more collective approach regarding energy policy and energy security (McGowan 2011; Trombetta 2012). Arguably, the Ukraine crisis has given a new boost to collective action to address energy security in the EU. The new EC that was installed in November 2014 even has a Vice-President for Energy

Union. Still, it is argued here that the completion of the internal gas system probably has a long-road ahead and that existing decision-making structures may not be sufficient to complete this journey. Hence in Chapter 7 several other components of the European gas system have been selected to complete the *ex ante* analysis. These are, in arbitrary sequence, the available and planned infrastructure capacities, the implementation of existing legislation, market trade and long-term contracts and the role of liquefied natural gas (LNG).

This book ends with Part IV (Chapter 8), which contains conclusions in terms of EU energy security and energy policy.

Notes

1   At the time of writing tensions between Russia and Ukraine continue, and it is widely believed that the failure to reach a cooperative modus could have an impact on the physical flow of commodity to the EU in the winter of 2015/2016.
2   See www.eurasiagroup.net/pages/top-risks-2015.
3   In January 2015, policy makers in Brussels are drafting concepts of what this Energy Union may entail. As these plans will evolve over the course of 2015, they are not part of this analysis. They are, however, reflected on briefly in the summary, which is written specifically for policy makers.
4   The debate on US exports of natural gas in the form of LNG is discussed in more detail in Chapter 7.
5   Starting with Gas Directive 98/30/EC.
6   See the considerations 9–12 of Electricity Directive 2009/72/EC, which state, for instance, that "without effective separation of networks from activities of generation and supply (effective unbundling), there is an inherent risk of discrimination not only in the operation of the network but also in the incentives for vertically integrated undertakings to invest adequately in their networks."

References

Ascari, S., 2011. *An American Model For the EU Gas Market?* Working paper RSCAS 2011/39. San Domenico di Fiesole: Robert Schuman Center for Advanced Studies, European University Institute.
Boersma, T., Johnson, C., 2012. The shale gas revolution: US and EU research and policy agendas. *Review of Policy Research* 29(4): 570–576.
De Vany A., Walls, D., 1994. Natural gas industry transformation, competitive institutions and the role of regulation. *Energy Policy* 22(2): 755–763.
European Commission, 2010. *Communication: Energy Infrastructure Priorities for 2020 and Beyond – A Blueprint for an Integrated European Energy Network.* COM(2010) 677 final. Brussels: European Commission.
European Commission, 2011. *Proposal for a Regulation on Guidelines for Trans-European Energy Infrastructure and Repealing Decision Number 1364/2006/EC.* COM(2011) 658 final. Brussels: European Commission. See http://eur-lex.europa.eu/LexUriServ/LexUriServ.do?uri=COM:2011:0658:FIN:EN:PDF.

European Commission, 2012. *Communication: Making the Internal Energy Market Work*. COM(2012) 663 final. Brussels: European Commission.

Harmsen, R., Jepma, C. J., 2011. North west European gas market: integrated already. *European Energy Review*, January 27.

Heather, P., 2012. *Continental European Gas Hubs: Are They Fit for Purpose?* Oxford: Oxford Institute for Energy Studies.

International Energy Agency, 2012a. *Golden Rules for a Golden Age of Gas*. World Energy Outlook Special Report on Unconventional Gas. Paris: OECD/IEA Publishing,

International Energy Agency, 2012b. *Medium-Term Gas Market Report 2012: Market Trends and Projections to 2017*. Paris: IEA.

International Energy Agency, 2013. *World Energy Outlook 2013*. Paris: IEA.

Le Coq, C., Paltseva, E., 2012. Assessing gas transit risks: Russia vs. the EU. *Energy Policy* 42: 642–650.

Makholm, J. D., 2012. *The Political Economy of Pipelines: A Century of Comparative Institutional Development*. Chicago, IL: University of Chicago Press.

McGowan, F., 2011. Putting energy insecurity into historical context: European responses to the energy crises of the 1970s and 2000s. *Geopolitics* 16: 486–511.

Neumann, A., Siliverstovs, B., 2005. *Convergence of European Spot Market Prices for Natural Gas? A Real-Time Analysis of Market Integration Using the Kalman Filter*. Dresden Discussion Paper in Economics no. 05/05. See http://hdl.handle.net/10419/22722.

Pearson, I., Zeniewski, P., Gracceva, F., Zastera, P., McGlade, C., Sorrell, S., Speirs, J., Thonhauser, G., Alecu, C., Eriksson, A., Toft, P., Schuetz, M., 2012. *Unconventional Gas: Potential Energy Market Impacts in the European Union*. JRC Scientific and Policy Reports. Brussels: European Commission.

Pelletier, C., Wortmann, J. C., 2009. A risk analysis for gas transport network planning expansion under regulatory uncertainty in Western Europe. *Energy Policy* 37(2): 721–732.

Renou-Maissant, P., 2012. Toward the integration of European natural gas markets: a time-varying approach. *Energy Policy* 51: 779–790.

Schmidt-Felzmann, A., 2011. EU member states' energy relations with Russia: conflicting approaches to securing natural gas supplies. *Geopolitics* 16: 574–599.

Smith Stegen, K., 2011. Deconstructing the "energy weapon": Russia's threat to Europe as case study. *Energy Policy* 39: 6505–6513.

Söderbergh, B., Jakobsson, K, Aleklett, K., 2009. European energy security: the future of Norwegian natural gas production. *Energy Policy* 37(12): 5037–5055.

Söderbergh, B., Jakobsson, K, Aleklett, K., 2010. European energy security: an analysis of future Russian natural gas production and exports. *Energy Policy* 38(12): 7827–7843.

Trembath, A., Jenkins, J., Nordhaus, T., Shellenberger, M., 2012. *Where the Shale Gas Revolution Came From*. Oakland, CA: Breakthrough Institute.

Trombetta, J., 2012. *European Energy Security Discourses and the Development of a Common Energy Policy*. Working paper no 2. Groningen: Energy Delta Gas Research.

Vazquez, M., Hallack, M., Glachant, J.-M., 2012. *Building Gas Markets: US versus EU, Market versus Market Model*. Working paper RSCAS 2012/10. San Domenico di Fiesole: Robert Schuman Center for Advanced Studies, European University Institute.

Von Hirschhausen C., 2008. Infrastructure, regulation, investment and security of supply: a case study of the restructured US natural gas market. *Utilities Policy* 16(1): 1–10.

Yergin, D., 1988. Energy security in the 1990s. *Foreign Affairs* 67(1): 110–132.

Part I
Framework of analysis

2 How to analyze European Union energy security

Introduction

This chapter introduces the framework for analysis that is used in this book. The first section gives an overview of energy security studies, a long-debated and often-disputed concept. This section demonstrates that most academic contributions that deal with energy security focus on energy markets (i.e. diversification of supplies, unreliable suppliers and transit risks). Other components of the energy system, such as infrastructure and regulatory authorities, are relatively under exposed. That mismatch is further explored in the third chapter of this book, which presents an analysis of the status quo of EU energy policy and explores the current decision-making and implementation structures within the EU.

The second and third section of this chapter elucidate on the theories that are applied on the case studies in the third part of this book. Successively these sections give a brief overview of neo-functionalism as one of the main streams in European integration theory and its links to energy security studies. This section also touches upon work from new institutional economics scholars, which is linked to this line of international relations (IR) studies. Subsequently this chapter outlines the concept of multilevel governance (MLG), which applies several components of neo-functionalist thinking. Although the merits of MLG as a theory have been fiercely debated, the conceptual framework provides a useful scheme to analyze the different case studies in the third part of this book.

Energy security studies

The debate on energy security has not passed unnoticed, but so far the only thing that stands out from the results is a lack of consensus about pretty much all aspects of energy security. As Chester (2010) concluded, energy security is a "wicked problem." First, two concepts are distinguished, namely energy security and security of supply. Some authors even use these terms interchangeably (Van der Linde et al. 2004; Kruyt et al. 2009). The remark of Kruyt et al. that these concepts are "synonyms" seems questionable, if only for the latter focuses on supply. Energy security on the other hand, at first glance seems to be open for more interpretations (e.g. necessary investments in infrastructure, technological development, regulatory challenges and stable demand). In the available literature there is a preference for the concept labeled energy security (when counting the number of references that is). Therefore, for the sake of clarity this book uses that concept as well. When referring to authors that have labeled the concept "security of supply" (e.g. Helm 2002; Chevalier 2006; Correljé and Van der Linde 2006), here the term energy security is used in order to avoid disorder.

The interest in energy security has been renewed following disruptions in Russian gas deliveries to the EU, and an increasing (perceived) pressure on global resources following the growth of in particular China and India. This has resulted in an outgrowth of interpretations of the concept of energy security, proposals for frameworks, conceptual considerations and admonitions and crisscross usage. At the beginning of this section it is appropriate to recap that proven global gas reserves and legitimate expectations about unconventional gas resources make that at current rate of consumption over 130 years of gas supplies are "likely" (Bothe and Lochner 2008: 22).[1] The IEA estimates that worldwide recoverable conventional gas reserves are around 400 trillion cubic meters, similar to global unconventional gas reserves. This number would at current rates of consumption be sufficient for 250 years of consumption.[2] In short, there is plenty of natural gas.

There is a wide variety of conceptual considerations about energy security in the literature. Clawson (1998) questioned the meaningfulness of the concept itself. Ciutâ (2010) expressed concerns about the compatibility of energy and security, for potentially coupling the two can result in a panoptic view. In other words, "energy security means the security of *everything*: resources, production plants, transportation networks, distribution outlets and even consumption patterns; *everywhere*: oilfields, pipelines, power plants, gas stations, homes; *against everything*: resource depletion, global warming, terrorism, 'them' and ourselves" (*ibid.*).

Scholars like Klare (2001) see energy resources as a cause or an instrument of conflict. However, empirical data on this causality between conflict and energy is generally scarce. Attempts to study terrorist attacks on energy infrastructures indicated that these attacks are "comparatively few" and that the "low percentage of attacks relative to other target types indicate that [energy infrastructures] are not a primary object of terrorist groups" (Toft *et al.* 2010). These findings nuanced a pledge that NATO must "play an increasing role in energy security" and "can provide an added value…in the area of physical protection of energy infrastructure…" (Tagarinski and Avizius, in Stec and Baraj 2009: 28). Considering the research by Toft *et al.*, this allocation of new tasks to NATO has at least a gleam of self-interest to it. Next to these concerns about the necessity to involve NATO in energy security, it is also worth considering the sheer reality of that mandate. With ten thousands of kilometers of pipelines, storage facilities, LNG routes, and production facilities, that seems overambitious.

Ciutâ (2010) referred to the usage of energy resources as a means of pressure, for instance in case of political conflict. An example of this is the gas supply disruption following the conflict between Ukraine and Gazprom in 2009 over payments in arrears, or at least "a big part of the problem was Naftogaz's failure to clear debts for gas delivered" (Pirani *et al.* 2009: 15). Smith Stegen (2011) concluded that these supply cut-offs must be attributed to economic causes and that over the course of history Russia more often than not failed to achieve political concessions using its energy resources as a "political weapon." Regardless, this dispute between Ukraine and Russia had far going consequences for citizens in countries like Bulgaria, urging the EU and national policy makers to propose additional regulation.[3] Högselius (2012) concluded that economic considerations have always been more important than political considerations in Russian–European energy relations. However, this does not mean that the "energy weapon" does not exist. He argued that its concept requires a broader view, which moves beyond supply disruptions, to include issues such as dumping natural gas on European markets, divide and rule strategies in which certain customers are favored over others, and so on (*ibid.*: 7).

Other scholars are more skeptical about supposed links between energy resources and conflict. Goldthau (2008) straightforwardly stated that energy weapons are "fiction," and that resources theoretically can be used as a political weapon only when all producers collectively decide to block supplies. According to him, the real challenges are in the lack of investments in Russia, to secure sufficient supplies for the longer term (see also Bothe and Lochner 2008; Söderbergh *et al.* 2010). While this can be true when upstream investments are concerned, investments in infrastructure (midstream), in particular to diversify supply routes to the EU (think of Nord Stream and former South Stream) are always perceived with suspicion and regularly lead to accusations of "geopolitical aggression (by Russia) against CIS countries and new Member States" (see Stern 2009).

An anecdote that supports the position of energy weapons being a fiction dates from the first oil crisis, when allegedly the Soviet Union – as one of the main oil producers – was requested by the OPEC members to cease its oil supplies to the states under embargo. One would have expected that ideological motives at that time would be decisive for the Soviets to make a decision to punish their capitalist counterparts across the Atlantic. However, with sky-rocketing oil prices on the world market they decided to do the opposite, namely to increase their sales to among others the US and the Netherlands (Goldman 2008). Of course also from this perspective the question is legitimate whether these examples exemplify a trend or are merely incidents.

Having said all this, there has been an ongoing debate about different natural gas tariffs that Gazprom charges its customers throughout the EU.

Many analysts argue that these are politically motivated (e.g. Smith 2006). The reality, however, is that European member states that suffer from higher tariffs for natural gas are the ones that do not have access to alternative natural gas supplies. Stern (in Henderson and Pirani 2014: 97) makes a convincing argument that Gazprom's pricing schemes appear to be designed as to maximize its commercial position in countries where alternatives are not available. This is the discriminatory behavior that one would expect from a monopolist, but that does not make the behavior "political."

All these considerations and conclusions fold into the theoretical debate whether energy resources have been securitized, and more importantly, whether that is helpful. Following the Copenhagen School of security studies, certain "existential threats" give legitimization for actions that are outside the normal political process (see Buzan *et al.* 1998). In other words, an energy supply disruption is considered to be an emergency and this validates the usage of any means necessary to restore it. There is ongoing debate whether this line of reasoning applies to the EU, in light of supply cut-offs in the 2000s. Trombetta appeared to halt between two opinions. She described the responses of the EC following the supply disruption in 2009 as a "securitization move" (Trombetta 2012: 21), but also acknowledged that it can be questioned whether this example represents a case of politicization rather than securitization (*ibid.*: 22). Özcan (2013) argued that the issue of energy is no longer a question of economics, but rather a matter of politics, and is now "finally" perceived in terms of an existential threat. However, his argument, like comparable studies, does not make clear what exactly securitization of energy resources has contributed to addressing the issues at hand.

Moreover, the case of the EU seems to provide ample evidence that market integration and an economic approach to energy resources deliver tangible results, whereas it remains unclear how securitization and grand rhetoric add value. Rather, they muddle the waters. McGowan (2011: 488) argued that disputes between Europe and Russia are better understood as being played out in a framework of politicization rather than securitization. In a comparison between the recent gas disputes and the 1970s oil crises he indicated that although in both instances governments declared a state of emergency, these moves focused on the allocation of resources and not a broader security agenda (*ibid.*: 493). It is worth noting, however, that opinions vary substantially across Europe on the question whether energy resources are securitized or merely politicized. Roth (2011) indicated that the Polish discourse regarding energy resources is highly securitized, with regular reference to notions of military security. A great example of this is the Polish (failed) attempt to initiate a European energy security treaty, along the lines of NATO (*ibid.*: 612). Other contributions have pointed at historical reasons for this divergence in approaches to reliance on Russia between eastern European member states and other member states (e.g. Schmidt-Felzmann 2011; Johnson and Boersma 2013).

One of the outcomes of the debate following the 2009 gas supply disruption has been the inclusion of "energy solidarity" in the Lisbon Treaty (see Roth 2011). Schmidt-Felzmann (2011) questioned to what extent this serves individual member states' interests or the European collective interest. Regardless, in terms of securitization, it is worth noting that nowhere in the Lisbon Treaty it is confirmed that a threat to energy solidarity has to be confronted by any means necessary. Rather, it is something member states strive for.[4] However, during the Ukraine crisis in 2014, there was a demonstrable will to use existing gas infrastructure to re-ship natural gas from eastern European member states – generally purchased from Gazprom – back to Ukraine.[5] Despite these signs of solidarity though, what remains is member states' individual sovereignty, as is explicitly mentioned in the Treaty. All this is in sharp contrast to, for example, the Carter doctrine, which left not much room for imagination.[6] In this statement, then President Carter of the US made clear that US oil interests in the Middle East would be protected if necessary, for "An attempt by any outside force to gain control of the Persian Gulf region will be regarded as an assault on the vital interests of the United States of America, and such an assault will be repelled by any means necessary, including military force." That fits the Copenhagen School definition of securitization much better than ongoing debates in Europe.

Diversification is another buzzword in European energy debates, and this concept too has a rhetorical component to it.[7] Different interpretations of diversification have been identified in relevant literature. First, some have called for diversification of supply routes through which natural gas is transported to the EU (Belkin 2008). From the onset, Russia has made it clear that initiatives to diversify away from the Russian supplies are not welcomed (Kalyuzhny, in Kalicki and Goldwyn 2005).[8] Russian authorities have actively undermined the European Union backed project of the Nabucco pipeline by closing deals with for instance Hungary and Serbia in order to make investment in this route financially unattractive and promote its own project South Stream (Belkin 2008).[9] Pickl and Wirl (2010) concluded that the overall economics of Nabucco, at least in terms of demand from gas shippers and cost-effectiveness, would have been positive. However, the question remained unanswered whether there were sufficient alternative suppliers next to Russia to make Nabucco a success – or to make an investment decision worth approximately €5 billion to start with. From this perspective it was not helpful that one major potential alternative supplier, Iran, was (and continues to be) excluded from economic activities. Furthermore, Monaghan (2007) argued that the Caspian Region poses serious security challenges in itself that must not be underestimated. The necessity of Nabucco has been linked to the proper functioning of the liberalized European gas system as a whole. According to Umbach (2010) without this pipeline market functioning in central and southeastern Europe "could hardly have been realized." However, in 2013, despite all the talk and political support for Nabucco, the operators of the Azeri Shah Deniz gas field opted for an alternative transportation route to Europe, namely one that crossed Greece and Albania into Italy.[10] This decision confirmed that at the end of the day volumes of natural gas consumed (Italy is the EU's second largest gas market, contrary to central and eastern Europe, where very modest volumes are consumed) are more important than political preferences.

The second form of diversification deals with energy suppliers. The EU seems rather diversified when it comes to natural gas. It consumes significant domestically produced resources and imports from in particular Russia, Norway and Algeria. Constantini *et al.* (2007) indicated that Europe is well located in the world market to benefit from increased exports of natural gas in the form of LNG from both Africa and the Middle East. Monaghan (2007) argued that shifts to other suppliers of natural gas would not necessarily reduce energy dependence, since most oil comes from those regions as well.[11] Cohen *et al.* (2011), in a study of OECD countries' energy security measured in diversification of oil and gas supplies, concluded that most countries have realized diversification of gas suppliers since 1990.

Despite these observations, political desires may incentivize the wish to use fewer resources from a certain supplier. In light of the 2014 Ukraine crisis, there have been widespread outcries throughout Europe to diversify away from Russian natural gas. Time will tell whether new policies are designed to curtail the share of Russian natural gas in the European fuel mix. However, initial empirical studies suggested that absent very drastic interventions in the traditional division of labor between public and private actors in the European gas market, a substantial reduction of the Russian market share is unlikely to happen (Boersma *et al.* 2014; Dickel *et al.* 2014). On the other hand, there are (two) examples of member states that have shown their willingness to pay a premium for natural gas from other destinations (i.e. Poland and Lithuania). These examples and their implications are discussed in more detail later in this book.

Some scholars cautioned that European initiatives to neglect Russia as a supplier would undermine Russian expectations that the EU will remain a reliable consumer in the future. This could result in a lack of investments and a Russian search for diversity away from European consumers (Monaghan 2007; Stern 2009). It could also frustrate key investments in exploration of new fields, infrastructure and interconnections between markets (Goldthau 2008). More broadly speaking, security of demand should be treated as an integral part of energy security (Cohen *et al.* 2011). In the end the attempts of the EU to move away from Russia could freeze the highly necessary investments in the Russian energy sector and hence turn the perceived negative dependence on Russian energy into a self-fulfilling prophecy. Banks (2007) suggested that the EU should consider lending funds to Russia to invest in exploration and transportation capacity, while collecting these loans over a longer period in the form of natural gas. In light of the crisis in Ukraine, however, at the time of writing closer collaboration between the EU and Russia seems unlikely. In fact, the spring of 2014 may well have created a context in which – after multiyear negotiations – China and Russia could finally reach an agreement on a 30-year supply contract, which is expected to commence in 2018.[12] Russian officials have announced that more agreements with China and other Asian countries are in the making, and time will tell whether they will actually come to fruition. At this point Russia, with its vast hydrocarbon reserves, could serve both the Asian and European markets, though ongoing sanctions against Russia do prohibit the development of new fields, in particular technically more challenging fields in the Arctic and unconventional oil production. Arguably, depending on how long these sanctions last and corporate investments in Russia diminishes, there could be a point in time where a lack of investments in green fields does prohibit Russia to serve both European and Asian demand.

The final form of diversification in the literature deals with energy resources. Again, the EU *is* rather diversified, looking at for instance the European energy mix. Next to considerable shares of oil and natural gas, coal-fired and nuclear plants each provide about one third of European electricity generation, alongside an increasing share of renewable energy (Belkin 2008). The EU's primary energy consumption is roughly defined along the same lines. It is important to note that natural gas is used not just for electricity generation, but most importantly as a feedstock for manufacturing and other industries, as well as heating and cooking.

Going back to the pile of literature on energy security, there are at least two (because of overlap somewhat arbitrary) clusters of academic focus. First, there are multiple contributions regarding the concept of energy security, in terms of what it means and/or what a definition could be, sometimes with specific reference to the EU (e.g. Bohi and Toman 1996; Van der Linde *et al.* 2004: ch. 2; Yergin 2006; Chevalier 2006; Belkin 2008; Kruyt *et al.* 2009; Hughes 2009; Chester 2010; Umbach 2010; Roth 2011; McGowan 2011; Trombetta 2012). In addition, there are interesting contributions about public perceptions of energy security, and discourse analysis in the US, EU, and Russia (Manley *et al.* 2013; Kratochvíl and Tichy 2013; Below 2013; Demski *et al.* 2014). Though this is a slightly different angle, these are included in the first cluster. The second cluster comprises contributions that focus more on EU energy policy, what policies are in place, and scenarios studies (e.g. Van der Linde *et al.* 2004; Correljé and Van der Linde 2006; Scheepers *et al.* 2006; Constantini *et al.* 2007; Von Hirschhausen 2008; Nuttall and Manz 2008; Pirani *et al.* 2009; Pointvogl 2009; Gasmi and Oviedo 2010; Söderbergh *et al.* 2010; Cohen *et al.* 2011; Proedrou 2012; Talus 2013; Aalto and Temel 2014; Helm 2014). On existing policies primary data are available, in terms of legislation, communications from European institutions, and occasional publications from European representatives (e.g. Oettinger 2010). The following sections contain a brief discussion of these two overarching clusters of literature on energy security.

To start with the first cluster, Bohi and Toman (1996) employ an economic viewpoint in their definition of energy security. They define energy security as "the loss of economic welfare that may occur as a result of the change in the price or availability of energy." They observed that over time "complex vulnerability issues" were added to the traditional connection of energy security with dependence on oil imports. This links closely to the "classical" definition of energy security, namely "the availability of sufficient supplies at affordable prices" (International Energy Agency 2001). Several authors added to this definition the notion that different countries have different interpretations of energy security (e.g. Yergin 2006; Chester 2010). Yergin (2006) argued that several occurrences such as power blackouts in the US, threats to attack energy infrastructure by al-Qaeda, concerns over the Iranian nuclear program, disruptions of gas supply in 2006 in Europe, growing resource nationalism, skirmishes in Nigeria and nature forces by the name of Katrina all gave sufficient cause for a fundamental revision of the traditional definition of energy security. It is worth considering that according to Toft *et al.* (2010) the terrorist attacks have not taken place on a relevant scale; forms of resource nationalism are found all over the world (Chapter 7 discusses the debate in the US about the export of LNG) and ecocatastrophes cannot be controlled. Hence, Yergin's argument can be nuanced to an extent. More importantly, even if that revision was necessary, it seems to have added just another definition to the list. What seems important from Yergin's argument is that context matters. Energy security from a European perspective is something very different than from a Russian perspective, where energy security is all about having stable markets to sell commodity (see also Kratochvíl and Tichy 2013).

Chevalier (2006) considered energy security to be "the whole physical and nonphysical supply chain." This includes reliable supply, transport and distribution of energy for a reasonable price. Time is an important dimension of the concept. Short term energy security can be threatened by political decisions, accidents and strikes, whereas structural political turmoil or insufficient investments can affect long term energy security (Chevalier 2006; Chester 2010; Söderbergh *et al.* 2010). According to Helm (2002) "supply can almost always be made equal to demand, provided the price is allowed to adjust. Only in extreme circumstances, such as embargoes, strikes or wars, is energy physically unobtainable."

Focusing on the European approach to energy security, Chevalier (2006) noted that the EC for several years favored to add uncertainties to the energy debate, such as climate change, geopolitical uncertainties, regulatory uncertainties and "the unexpected." Helm (2014) noted that European energy and climate policies are going "nowhere," and contribute to driving down competitiveness, driving up prices, and not making much difference for climate change. Umbach (2010) argued that the European approach towards energy security had to be drastically changed, following the supply disruptions of 2006 and 2009. Several authors concluded that one approach to energy security ought to be adopted and acted upon by all levels of government involved (Chevalier 2006: 18; see also Le Coq and Paltseva 2012: 648). However, at the same time there is substantial agreement that the concept changes over time (Yergin 2006; Chevalier 2006; Chester 2010; Kratochvíl and Tichy 2013). It is questionable whether including all the mentioned themes under one umbrella can actually help to solve the problem. This is especially the case since this problem appears to be redefined constantly, and differ depending on what perspective you have. In addition, there is no consensus about which challenges and uncertainties should be included when studying energy security. So far stable and sufficient supplies for reasonable prices are about the only elements in the different definitions of energy security that have endured the course of time. Given the lack of consensus among people that study or work on the matter, it can be no surprise that energy security is not more than an emerging public discourse. Based on the limited available empirical work, people's concerns revolve around dependence, and environmental concerns (e.g. Manley *et al.* 2013; Demski *et al.* 2014).

One thing that stands out from the available literature is (in light of the aforementioned four facets of the EU energy system) that there appears to be a lot of attention for both markets and governmental institutions. Chester (2010) confirms this observation, reporting "an almost overwhelming focus on securing supplies of primary energy sources and geopolitics." On occasion infrastructure and regulation have been mentioned. Umbach (2010) referred to massive investment programs in infrastructure that are necessary in the next decades ahead. Helm (2002) and Chevalier (2006) mentioned network security as one part of energy security. Some have identified the incomplete liberalization of the EU energy system as a risk (Chevalier 2006; Helm 2014). Others blamed liberalization to be the driving force in delayed investments in infrastructure and hence pose a threat to energy security (Meyer 2003). Several studies have also looked at energy infrastructure from a military perspective (Yergin 2006; Tagarinski and Avizius, in Stec and Baraj 2009; Toft *et al.* 2010). Yet only a few studies focus on gas infrastructure and regulation as being part of a well-functioning energy system and contributing explicitly to energy security. Jamasb and Pollitt (2008) suggested that energy security can be strengthened by regulation that adequately stimulates investment in necessary infrastructure, albeit pipelines, storage facilities or interconnection capacity. This leaves a crucial task for the regulatory authority, which searches for a balance between ensuring investment and reasonable tariffs from a consumer's perspective. Jamasb and Pollitt focused on the electricity market, although the stimulation of investments in gas storage facilities in Belgium and the United Kingdom is touched upon (*ibid.*: 4586). As is discussed in more detail in Chapter 5 of this book, there is a wide range of literature on the regulation of natural gas infrastructure, focusing on sufficient infrastructure investment in a liberalized gas system (e.g. Von Hirschhausen 2008; Gasmi and Oviedo 2010). These analyses, however, focus more on energy economics, and references to energy security are scarce.

Chapter 3 demonstrates that the EC has (partly) broadened its focus from securing supplies to also include gas infrastructure and regulation, but in the academic literature the importance of gas infrastructure or infrastructure in general is frequently "overlooked" (Jamasb and Pollitt 2008; Hughes 2009). This is confirmed also by the literature overview in Table 2.1, which aims to indicate on what parts of the energy system aforementioned contributions predominantly focus.

In sum, energy security is and remains a contested concept. One element that returns in all contributions is the sufficient availability of affordable energy supplies. Another point of consensus seems to be the change of the concept over time and the different interpretation it receives in different circumstances. What can be confusing for observers is the oft-quoted terminology of "markets" in energy related studies. In several cases it is not explicitly mentioned what part of the energy system (markets, infrastructure, regulatory authorities, government institutions) is dealt with. The discussion furthermore suggests that the framework of security does not fit energy as such. Even though it is often used, in practically all cases it is questionable whether the issue has in fact been securitized or merely politicized. In an analysis of discourses used both by EU and Russian representatives, Kratochvíl and Tichy (2013) concluded that the discourse of integration (focusing on mutual benefits and the complementarity of both parties) is dominant both on the Russian and the European side. Thus, they found claims about energy issues being securitized exaggerated. What also stands out is the relatively modest attention that is given to two crucial elements of well-functioning energy systems, namely infrastructure and regulatory authorities. There are abundant contributions on infrastructure that take a more regulatory economics perspective (as is discussed in Chapter 5) but those contributions generally do not focus on energy security. The academic focus on the availability of energy supplies is especially interesting given the fact that natural gas is abundantly available. All this leaves the question open whether supplies, although at the center of the academic debate on energy security, may in fact not be the most acute challenge for the EU. This is considered in Chapter 8 of this book.

The next section turns to the following building block of the theoretical framework, namely (neo)functionalism and new institutional economics.

(Neo)functionalism and new institutional economics

Functionalism as a theory of international relations arose parallel to the development of regional and global interdependence. Interdependence

Table 2.1 Overview of the main focus of the examined theoretical contributions on energy security

Reference	EU energy system				
	Markets	Infrastructure	Regulatory authorities	Governmental institutions	Definition or concept of energy security
Aalto and Temel	X			X	X
Belkin	X			X	
Bohi and Toman	X				
Bothe and Lochner		X			
Chester					X
Chevalier	X	X			X
Ciutâ					X
Clawson					X
Cohen *et al.*	X				
Constantini *et al.*	X	X			
Correljé and Van der Linde	X				
Dehousse	X				
Gasmi and Oviedo		X	X		
Goldthau	X	X			
Helm	X	X	X	X	X
Hughes	X	X		X	
Jamasb and Pollitt		X	X		
Kalicki and Goldwyn	X			X	
Kruyt *et al.*					X
McGowan				X	X
Meyer	X			X	
Milina	X			X	
Monaghan	X				
Nuttall and Manz				X	
Pickl and Wirl		X			
Pirani *et al.*	X				
Pointvogl	X			X	X
Proedrou	X			X	X
Roth				X	X
Scheepers *et al.*	X			X	
Söderbergh *et al.*	X				
Stec and Baraj				X	
Stern	X				
Talus	X	X	X	X	
Toft *et al.*		X			
Trombetta				X	X
Umbach	X				X
Van der Linde *et al.*	X				X
Von Hirschhausen		X	X		
Yergin	X				X

appeared to make issues complex, given for instance the broad variety of cultural and ideological differences. Functionalism on the other hand aimed to dismantle these complex issues until the technical aspects remained, in order to be able to organize issues at a certain required scale. The predecessor of the EU, the European Coal and Steel Community, was even defended politically with rhetoric based on functionalism (De Wilde 1991). Also, its successor neo-functionalism and the founding architects of the European Community have also been connected to each other (Rosamond 2000: 50). To Mitrany (1965), who was the founding father of functionalism, the continuous development of common activities and interests across territorial borders was the only way to make those borders irrelevant.

Functionalism encircles two basic principles, namely form follows function and spill-over (*ibid.*). The first principle entailed the conclusion that there is no such thing as a blueprint to solve an issue. Rather, the characteristics of that issue had to determine the approach to solve the problem at hand, instead of political interests, for example. Following this reasoning some problem would be best served by ignoring the conventions of national territory (Rosamond 2000: 33). This distinction between technical and political aspects of an issue has been criticized for its rather subjective nature. In addition, one could question how realistic it is to dispose an issue of its ideological and political content, when all people and institutions involved actually carry and maybe even *form* that very same content. De Wilde (1991) suggested that an enrichment of functionalism would be to specify for whom something is functional and to what purpose.

The second principle of functionalism, called spill-over, deals with the expectation that cooperation on technical issues between states would have a de-escalating effect on power politics between those states. In other words, if actor A can collaborate with actor B on certain issue X, why would they slaughter each other over other issues? To Mitrany (1975), these joint functional arrangements formed the only serious roadmap towards more stable and peaceful relations between states. Hence what functionalism shared with interdependence theory is the notion that the causes of war and the conditions for peace are located in the structure of society. The reduction of chances of war, however, was a bonus of the functionalist approach. Basically leaving the nation state as a given boundary and providing transnational institutions with the opportunity to serve as providers for welfare, had two effects. The first effect was loyalty transfer away from the nation state and the second a bonus of reduction of the chances of international conflict (Rosamond 2000: 33). A question unanswered is what this would mean for the concept of a state as such. According to De Wilde (1991) many scholars criticized functionalism on this issue and Mitrany also seemed ambiguous, sometimes referring to broad functional political organizations but also to massive centralization in national government. One could question whether in fact there was a choice to be made here, or whether different levels of government and governance could exist simultaneously. Yet throughout his work Mitrany also opposed regional integration arrangements, for regions would lead to the same faults of the state system, only at another scale (Rosamond 2000: 37). Following function instead of form, Mitrany would probably reason that incorporating Norway and Russia into some sort of EU energy cooperation makes more sense than excluding them. This line of reasoning also formed the basis of the Energy Charter Treaty, which aimed to include Russia as a major external supplier of European resources. Yet the very notion of a "union" constitutes a form – that is indeed not rigid and definitive – over function instead of the opposite.

In response to Mitrany and other functionalists, scholars like Schmitter (1969) began to doubt whether functional needs alone could in fact be sufficient to push forward regional integration. Functionalism suggested that integration would be steered by rationally established needs and a technocratic process would do the trick, but in fact that line of reasoning would rule out "the political" (Rosamond 2000: 40). These so-called neo-functionalists argued that regional integration needed more: first, functional spill-over can take place only when integration happened in a functionally related area. Second, there is pressure on the members of the collaboration in order to adopt a single policy, a process called externalization. Finally there is politicization, meaning the process by which regional integration is challenged among a widening circle of political actors (Schmitter 1969: 161). Hence a major distinction from functionalism was that neo-functionalists added to the technocratic process as described by Mitrany the active steering of the process itself by participants by "pursuing their own self-interest" (Rosamond 2000: 55). But the line of reasoning of neo-functionalists went further. Next to advocating the advantages of integration, the newly built transnational institutions were also expected to be "entrepreneurial" (*ibid*.: 58), a process which has also been labeled "cultivated spill-over" (Tranholm-Mikkelsen 1991). Subsequently, within the nation state, interest groups would experience the benefits of integration, accordingly acting positive towards their own national governments. Hence the process would get another stimulus from within the nation state, next to the catalyst function the transnational institutes fulfilled. It could be questioned though whether there is a limit on the willingness of national actors to promote spill-over. At a certain stage it would result in the complete transfer of power to another level, in this case that of the EU, hence undermining the influence and potentially the legitimacy of the national actor involved. As examined in more detail in the third chapter, a more European approach towards energy regulation in the form of the Agency for the Cooperation of Energy Regulators (ACER) was initiated in 2010. This potentially undermines the influence of national regulatory authorities by opening the door towards European regulation, though this is a lengthy process, as this book describes later.[13] For the time being, Helm (2014) concluded that the time-consuming trajectory towards one internal market causes more problems than gains for the EU.

Neo-functionalists had a major problem with the mechanisms that would steer transfer of loyalty to a transnational political community. To them this was mainly a technocratic process, whereas daily practice demonstrated that nationalism and/or politics could play a key role in the integration process.[14] This is where most prominent integration theories appear to struggle. All seem to "pick sides" explicitly. Neo-functionalists and supra-nationalists struggle with national influences and decisions to guide national sovereignty. Intergovernmentalists have difficulties with the increasing role of European institutions and in particular their own initiatives, which sometimes run counter to national interests. In this theoretical approach integration is often reduced to a single decision to integrate or not. However, the process might deserve more attention, or "integration must be conceived of as a process of action (decision to integrate) and reaction (response to integration) ... since integration proceeds in stages, the dialectics of the process has to be given more attention" (Corbey 1995).

Corbey discussed what she labeled "dialectical functionalism" as a framework to analyze European integration. It is an amendment to neo-functionalism, for her approach does take the response to integration into account. The debated spill-over effect takes place and generally leads to reverse movement by member states (i.e. safeguarding adjacent policy areas against EU intervention; *ibid.*: 263). In a nutshell the functional linkage suggested by neo-functionalists from this perspective is located at the *national* level. Integration in a certain policy area results in increased national government intervention in adjacent policy areas. This leads to policy competition between member states, a movement that in the end can prove to be counterproductive. To give an example, it is worth considering the various speeds of implementation of the so-called Third Package of European legislation. A solution to break out of that status quo might be to transfer policy responsibility to the European level. National interests, however, so far hinder alternatives to develop and a significant number of member states yet have to implement legislation that should have been implemented in the spring of 2011. This issue is discussed in more detail in Chapter 7.

So what to deduce from this brief overview? First, Mitrany's functionalism runs into political walls from the very beginning. European policy makers have opted for close cooperation on energy matters with important external suppliers. But in particular regarding Russia the relationship remains fragile, which for instance becomes clear when considering the Third Package and its reciprocity clause.[15] Boussena and Locatelli (2013) argued that this institutional divergence increasingly drifts Europe and Russia apart. Apparently both sides are taking different courses, despite the fact that both arguably would benefit from stable long-term mutual relations (in terms of secure supplies of natural gas and stable demand for it). Bilateral relations have further deteriorated over the war in Ukraine. At the time of writing, it is highly uncertain what the long-term effects of this conflict will be for Russian–European relations.

Second, neo-functionalists have struggled with the question who is actually steering the European integration process. As is discussed in more detail in Chapter 3, both European institutions and member states are in the driving seat when it comes to energy policy. European institutions have drafted ambitious statements from 1955 and onward, policy initiatives dating from the first oil crisis, and energy directives from the 1990s onward. At the same time, there is a degree of reluctance to implement this European legislation within several member states. An example is the hesitance in some member states regarding ownership unbundling, one of the crown jewels of the liberalization of European energy markets. It is likely that national interests of both French and German integrated energy companies made sure that the Third Energy Package contained an acceptable alternative to full ownership unbundling, that was in fact desired by the EC (and already put in place by several other member states, subsequently disturbing the European "level playing field"). The EC has since been active in reaching its objective through a detour, as was for instance shown by former Euro Commissioner of Competition Kroes. She made deals with German integrated energy companies RWE and EON to renounce a part of their networks, respectively its gas network Thyssengas and its electricity network Transpower, after allegations of abuse of market power. In the case of Germany, national policies have also played an important and probably decisive role in the integrated companies eventually splitting up. Under pressure of the nuclear phase-out in Germany and the accelerated development of renewable energy, German company E.ON announced in November 2014 that it would split up its company.[16]

By now the EC has shifted its attention to energy infrastructure and regulation, in its repeated calls for completion of the internal energy system (European Commission 2011, 2012). The ongoing debate within the EU about the role of the EC with regard to energy infrastructure and in particular financing of that infrastructure is fascinating. On the one hand many of these infrastructural projects are currently delayed or not built (an overview is presented in Chapter 7). This is due to modest volumes of gas consumption, different regulatory regimes, changing tariffs, costs and benefits that are not shared by member states and so on. There are companies and countries involved that would like the EC to participate and help solve these issues and there are those, like the Netherlands, that are highly skeptical of any role of the EC in this matter. The argument is as follows: the Dutch transmission system operator Gas Transport Services has made major investments throughout the years in order to position itself in a favorable position to turn the Netherlands into a "gas roundabout" (in short: a crossroads of gas flows through northwestern Europe). If with financial support from the EC alternatives such as a Belgian roundabout would be financed that is seen as a distortion of competition.[17] The point here is that member states based on their own interests or beliefs, correct and subsequently steer energy policy initiatives of the EC. Whereupon daily practice demonstrates that the EC sometimes responds by altering its course, for instance when initiatives aimed at better market functioning (full ownership unbundling) stagnate and the EC subsequently focuses on other parts of the energy system instead (i.e. infrastructure and regulation). This example confirms that it is important that an analysis of regional integration should also place substantial emphasis on the role of non-state actors in providing the dynamic for further regional integration.

One of the fundamental issues in European integration theory is who or what is in fact steering the process of integration. The book aims to demonstrate that the status quo brings additional risks in terms of European energy security and that particular risks that are easily overlooked come forth from existing institutional flaws. The importance of institutions for the success or failure of economic activity has been broadly acknowledged by scholars in the field of what has been called new institutional economics. The term "new" is mainly used to distinguish between earlier work on institutions, and current work, the main difference being that institutions are now perceived to be susceptible to analysis (Williamson 1998). In short, economists study the interplay of demand and supply and how that determines prices but neglect the context in which this interplay takes place (see Coase 1998). Yet together with that standard constraint of economics institutions in fact are crucial in determining transaction costs, production and hence the feasibility of engaging in economic activity (North 1991). Institutions in this context are both informal constraints (e.g. sanctions, taboos, customs, traditions, codes of conduct) and formal rules (constructions, laws, property rights) (*ibid.*: 97). Generally speaking major changes in the rules of the game (formal rules) occur on the order to decades or centuries, with the occasional exception of a sharp break of established procedures following, for example, civil wars, perceived threats, financial crisis or perceived threats (Williamson 2000). From that perspective one could argue that the gas supply disruptions in 2006 and 2009 were not large enough to incentivize institutional change deemed necessary to make the EU internal gas system work. If they had been significant enough, more efforts would have been made to further integrate markets. Without solid institutional foundations a gas system cannot perform properly. It therefore makes sense to analyze existing rules in the EU gas system and establish whether the "formal rules of the game are right" (*ibid.*: 595).

Figure 2.1 presents a schematic overview of new institutional economics. The "formal rules" (North 1991) are depicted at level 2, where instruments include "executive, legislative, judicial and bureaucratic functions of government as well as the distribution of powers across different levels of government" (Williamson 2000). Also of importance to this analysis of the EU gas system are governance structures, pictured at level 3. As the subsequent chapters demonstrate, it is at these levels of analysis where shifts

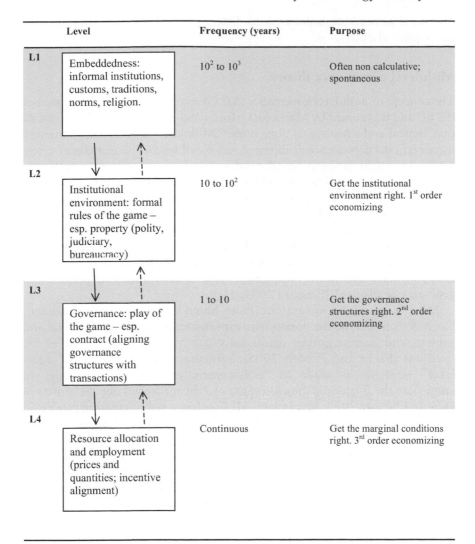

Level	Frequency (years)	Purpose
L1 Embeddedness: informal institutions, customs, traditions, norms, religion.	10^2 to 10^3	Often non calculative; spontaneous
L2 Institutional environment: formal rules of the game – esp. property (polity, judiciary, bureaucracy)	10 to 10^2	Get the institutional environment right. 1st order economizing
L3 Governance: play of the game – esp. contract (aligning governance structures with transactions)	1 to 10	Get the governance structures right. 2nd order economizing
L4 Resource allocation and employment (prices and quantities; incentive alignment)	Continuous	Get the marginal conditions right. 3rd order economizing

Figure 2.1 Economics of institutions

Source: derived from Williamson (2000: 597)

between governance levels, compromises, and subsequent vagueness prevail.

The case studies outline what responses follow from integration in certain policy areas, next to the functional shift within decision-making structures. The next section of this chapter explores the options to use a multilevel governance framework to dissect the different case studies in the subsequent chapters.

Multilevel governance theory

The concept of multilevel governance (MLG) resulted from the appearance of the EU and was coined by Marks (in Cafruny and Rosenthal 1993). It consists of a vertical and a horizontal dimension. "Multilevel" refers to the witnessed increased interdependence at different territorial levels. "Governance" refers to the identified growing interdependence between government and non-government actors at different territorial levels. The concept of governance is much broader than the more traditional government. The latter leans on more formal mechanisms such as sovereignty and constitutional legitimacy, while governance in addition rests on informal agreements, shared premises and successful negotiations. In other words, governance refers to "any collectivity, private or public, that employs informal as well as formal steering mechanisms to make demands, frame goals, issue directives, pursue policies and generate compliance" (Rosenau 2004: 31).

MLG is not about integration (more about explaining policy making), but it does underline the notion that supranational actors, both public and private, and interest groups significantly contribute to the shaping of EC decisions (Bache and Flinders 2005). Sometimes next to the term "multi-level" reference is made to "poly-centric governance," in order to emphasize the functional dimension next to the territorial one (Schmitter in Diez and Wiener 2003: 49). In fact, MLG offers a polity-creating process in which authority and influence to make policies are shared across multiple levels of government (Hooghe and Marks 2001). This contrasts with more state-centric theories in which governments are the ultimate decision makers that transfer only limited power to supranational institutions to achieve specific goals. These governments from this perspective are nested in autonomous national arenas from where they determine EU policy making. In MLG, leaving from the starting point that there is no monopoly on the decision-making competencies in the EU, in order to explain policy making it is required to analyze the independent role of European actors such as the EC and the EP. In addition Hooghe and Marks (*ibid.*) argued that through the process of collective decision making follows self-evidently a loss of control for individual member states. Instead of political arenas being nested, they are interconnected, hence there is no clear separation between national and international politics. One could wonder why political actors would actively weaken institutions that they work for at a particular moment. Hooghe and Marks (*ibid.*) argue that this is because these actors follow their own normative and private preferences. This observation on the two-level game between domestic and international interaction has been supported in other theoretical contributions. The complexity is caused since rational behavior at one policy level can be impolitic for that same player on another policy level (Putnam in Lipson and Cohen 1999). This apparent link between national and international politics resembles a concept called "linkage politics" (Rosenau 1969). To give an example, a certain topic might be too delicate for the domestic political arena and is subsequently transferred to another level of policy making. In addition that political actor could be searching for re-election and hence behave rather opportunistic at a certain moment.

Over time the concept of MLG has been refined into type 1 and type 2 MLG (Hooghe and Marks 2003). The first is dominant among scholars who witness a modification of the traditional state and who recognize more intense transnational relations, an increase in the importance of public and private cooperation and hence in addition an increased importance of multinational companies. These scholars also retain the central role of the nation state. This type of MLG has originated from federalism in which a limited number of governments operate separately at different territorial levels. It suggests a certain hierarchy. These governments are the unit of analysis. Type 2 MLG refers to "task-specific jurisdictions" in which scholars focus on a particular policy problem. In a nutshell citizens are not served by "the" government, but by a range of different public service industries that form functional associations. These industries do not form a tight fit, but partly overlap. Both types of MLG are assumed to complement each other.

Many scholars have criticized (parts of) the concept of MLG from its start and there appears to be no consensus regarding its utility so far, as illustrated by Bache and Flinders (2005). Rosenau (in *ibid.*) attributed certain use to MLG and acknowledged governance as a broader concept than government, which follows from the growth of complex interdependencies as a result of which rules are established in all sorts of non-governmental organizations as well. In other words, "states are no longer the only players."

Another important issue following from MLG is what the role of the traditional state is or can become. Within MLG the state is one of the kids on the block, whereas its exact position in the hierarchy remains vague. In more state-centric approaches the EU is viewed as the appearance of a new supranational arena in which states attempt to pursue their national interests (Jessop in Bache and Flinders 2005). In this approach the state in fact remains "the" kid on the block. The shift of power to the supranational level, paradoxically orchestrated by national states themselves according to these scholars, can in the end result into an up-scaling of the traditional state to a supranational statehood. The process in between in combination with the increased complexity of relevant actors and the absence of legal frameworks within MLG is actually risky and a threat in terms of democratic accountability (Papadopoulos 2007). Moravcsik (1994) noted that some scholars tend to exaggerate the changes in intergovernmental relations and underestimate how easy states can strengthen their grip on a political process if considered necessary. In addition member states of the EU can use

MLG as a means of enhancing their independence from societal actors and can hence increase their steering capacity (*ibid.*). MLG assumes that including more actors in the policy process does not have influence on the joint capacity to reach decisions, which is doubtful. Also negotiations within MLG are sometimes assumed to be non-conflicting, but how realistic is this? Peters and Pierre (in Bache and Flinders 2005) suspected that the lack of formal means of decision making could be a pitfall and warned that MLG could prove to be a "Faustian bargain."

One of the difficulties of MLG is its coordination dilemma: policies of one jurisdiction have spill-overs or externalities for other jurisdictions, which require coordination (Hooghe and Marks 2003). Since all those involved are aware of this, it results in free-riding behavior being a dominant strategy of large groups within MLG systems. To limit this behavior one can reduce the number of autonomous actors (in fact type 1 MLG) or the interaction among them, which seems more difficult. It is questionable whether this dilemma is a persistent problem of MLG or more a temporary rigidity that comes with the process of relocating decisional powers from a central level to multiple central levels or creating new powers in the process. The latter option opposes one of the claims by intergovernmentalists that states pool power in the EC arena. Matláry (1993) has argued, in a study on European energy policy development, that convergence between member states and European institutions' interests can be created largely by the EC, and hence new power can be as well.

Away from the theoretical considerations MLG has been applied on various occasions. George (in Bache and Flinders 2005) believes that MLG has the strength of a theory in it and that it offers a way to explain what sort of an organization the EU is: central executives of state partly do the governing, but in addition share responsibility with other actors, both supranational and sub national. Others have serious reservations about the value of MLG as a theory and refer to the logic sharing of control over activities rather than monopolizing it by national states. Fairbrass and Jordan (in Bache and Flinders 2005) have identified environmental policy as a "case par excellence of the dispersion of authoritative decision making across multiple territorial levels." They concluded that the essence lies in the fact that states do not watch passively when sharing control. Aalto and Temel (2014) used the English School of IR and concluded that although the EC is the leading formal institution advocating gas market integration in the EU, it is hindered by the slowness of many member states to implement existing legislation, and the opposition to its policies by some companies. Furthermore the EC is constrained by in particular Russia's adherence to state capitalism, which incentivizes several member states to pursue hedging strategies through bilateral deals (*ibid.*: 9).

Several studies with MLG frameworks focused on climate policy (De Bruijn 2002; Monni and Raes 2008) and renewable energy policy (Rabe 2007; Hirschl 2009). These studies too are far from unambiguous about how to apply MLG. According to Rabe (2007) climate change as a subject of study is well suited for MLG since national governments and classic international relations are not sufficient for this challenge. Whereas MLG was originated to study policy making within the EU, in this effort MLG is applied to the US and Canada both federal and state/provincial governments in order to expose the institutional capacities of (sub)national governments to develop and implement policies that help stabilize or even reduce carbon emissions. One of the assumptions here is that experience in MLG systems proves that an international accord in terms of carbon emission reductions does not automatically lead to these reductions. This analysis demonstrates that tangible results have been booked in particular on the state level in the US, whereas the national government has failed to deliver. The Canadian case demonstrates exactly the opposite. There is some evidence that supports the notion that government policies from the local to the federal level, including renewable portfolio standards, energy efficiency requirements and car mileage usage, have generated results in the US (Delaquil *et al.* 2012). In contrast with the suggestion put forward by Rabe, others have argued that unilateral actions are not very helpful to address the problem at hand (De Bruijn 2002). Others warned that sub national initiatives are subject to potential free rider issues (Monni and Raes 2008). De Bruijn argued that these types of policy consist of a continued process of bargaining between levels of governance.

In sum, there is no consensus regarding the usefulness of MLG as a theory. The critique that MLG lacks formal means of decision making raises questions, for within the EU a continued process of shifting of decision-making power between various territorial levels can be observed, as expressed when using MLG as a scheme for analysis. Providing insight into these dynamic and complex processes is one of the major advantages of using MLG as a research tool. Maybe in addition to this policy-making process, *implementation* of these policies and realization of them is a display of this process of shifting as well, yet these two processes do not need to be symmetrical. To give an example, European member states have been rather progressive in designing and accepting several guidelines concerning climate change and renewable energy (Trombetta 2012). MLG provides a useful scheme to analyze this process as well, as is demonstrated by a study of climate policy initiatives in Finland by Monni and Raes (2008).

Second, there are several cases in which MLG has been applied as a concept in different research settings, both within and outside the EU. The latter in itself can raise new questions, while the US and Canada have different domestic dynamics in terms of decision-making processes. Despite these uncertainties, MLG offers an elaborate framework to analyze policy-making processes and decision-making structures in the EU by distinguishing between several territorial levels and in addition providing a difference between public and private actors. Also, there is some experience with the application of MLG in other parts of the world, notably the US.

Therefore MLG can contribute to the dissection of the case studies in Chapters 5–7, in order to search for an answer whether existing decision-making structures in the EU energy system affect European energy security.

Discussion

Though it is virtually impossible to review all academic contributions on energy security, the overview presented in the first section of this chapter confirmed the earlier assumption that most contributions to the debate focus on markets (albeit the availability of supplies, supply disruptions or unreliable suppliers). Less has been written about other vital components of the energy system (i.e. infrastructure and regulation), in particular regarding the internal market. Hence, the analysis confirms the necessity for a broad analysis of the EU energy system, a contribution that this book makes.

The case studies in the next chapters are expected to provide a good test for the limits to (neo)functionalism as presented in this chapter, in terms of spill-over, how it relates to energy policy in the European Union, and in terms of the other basic principle – form follows function – for example, in relation to the oft-quoted collaboration between Russia and the EU. Following Rosamond, there are signs of certain "entrepreneurship" in European institutions regarding energy resources. New institutional economists suggest that the study of functions of government and the distribution of power across levels of government are crucial to understand whether the preconditions to proper market functioning are in place. Given the broad acknowledgement that the US gas system currently is the only well-functioning gas system in the world, it makes sense to use that system as a benchmark in the analysis of the European gas system, and examine whether lessons can be learned. Multilevel governance theory may be debated as a theory per se, but it does provide a useful framework for analysis. In addition, it has been applied to the US in the past. Therefore these building blocks supplement each other and form a framework of analysis.

Notes

1   The term "likely" has been translated from the German "*wahrscheinlich*."
2   See www.iea.org/aboutus/faqs/gas (accessed March 31, 2013).
3   For instance Regulation 994/2010 on security of gas supply, which is discussed in more detail in Chapter 4.
4   Lisbon Treaty, article 176A sub 1.
5   Because of an ongoing pricing dispute between Ukraine and Gazprom, supplies to Ukraine were eventually halted in June 2014. Before that moment, and in light of the fear of supply disruptions, several European member states (i.e. Poland, Slovakia, and Hungary), expressed their willingness to re-ship natural gas to Ukraine, in order to make it more resilient against supply shocks, even though it is generally assumed that this would not be sufficient to counter a supply disruption in the winter of 2014/2015. See www.economist.com/

blogs/easternapproaches/2014/04/slovak-ukrainian-gas-deal (accessed October 28, 2014).
6   For a blog reference, see www.worldaffairsjournal.org/blog/andrew-j-bace-vich/carter-doctrine-30.
7   The idea to diversify the EU energy supply dates from the 1991 Energy Charter Declaration, which was followed in 1994 by the Energy Charter Treaty (legal status acquired in 1998). Arguably the most evident example of this EU strategy has been the Nabucco pipeline, which eventually was not built because the market preferred another route.
8   For the relevant interview, visit www.euractiv.com/en/energy/eu-wrong-priori-tise-energy-diversification/article-176380.
9   The Nabucco pipeline intends to bring natural gas from the Caspian Region to Europe without transiting Russian territory. At the end of 2008 Russia closed its deal with Serbia (*Financial Times*, December 24, 2008).
10 See http://europa.eu/rapid/press-release_IP-13-623_en.htm; although the decision was publicly supported by the EC, it was disappointing that the new transport route did not bypass countries in central and eastern Europe that so direly needed access to alternative gas supplies.
11 Russia possesses the most conventional resources of natural gas, followed by Iran, Qatar, Saudi Arabia, and the United Arab Emirates.
12 For a short discussion on this deal, visit www.globaltimes.cn/content/863631.shtml.
13 See Regulation 713/2009, for instance article 8, sub. 1 a and b, stating conditions under which the Agency for Cooperation of Energy Regulators decides upon regulatory issues that fall within the competence of national regulatory authorities, which may include the terms and conditions for access and operational security.
14 Rosamond refers to the entry of the French president Charles de Gaulle (2000: 67), but there are numerous nationalistically oriented movements of national governments throughout the EU that criticize further integration, although arguably with the purpose of in fact enhancing it.
15 Directive 2009/73/EC concerning common rules for the internal market in natural gas, as part of the Third Package, contains a reciprocity clause that was introduced into the third liberalization package in order to avoid indiscriminate acquisition of EU energy grids by third countries. This clause is widely regarded as targeting mainly Russian Gazprom.
16 See http://uk.reuters.com/article/2014/11/30/uk-e-on-divestiture-idUKKCN0JE0TZ20141130.
17 This arbitrary financing becomes clear when examining the European Economic Recovery Plan and its outcomes, which demonstrate that the financial means have been distributed across the EU, while some parts of the Union clearly could lay claim to the majority of those funds, because that is where objectively most investments are required. For more information on the EERP, see http://ec.europa.eu/economy_finance/publications/publication13504_en.pdf.

References

Aalto, P., Temel, D. K., 2014. European energy security: natural gas and the integration process. *Journal of Common Market Studies* 52(4): 758–774.
Bache, I., Flinders, M., (eds) 2005. *Multi-Level Governance.* Oxford: Oxford University Press.
Banks, F. E., 2007. *The Political Economy of World Energy. An Introductory*

Textbook. Singapore: World Scientific Publishing.

Belkin, P., 2008. The European Union's energy security challenges. *Connections* 7(1): 76–102.

Below, A., 2013. Obstacles in energy security: an analysis of congressional and presidential framing in the United States. *Energy Policy* 62: 860–868.

Boersma, T., Mitrova, T., Greving, G., Galkina, A., 2014. *Business As Usual: European Gas Market Functioning in Times of Turmoil and Increasing Import Dependence*. ESI policy brief 14-05. Washington, DC: The Brookings Institution.

Bohi, D. R., Toman, M. A., 1996. *The Economics of Energy Security*. Boston, MA: Kluwer Academic Publishers.

Bothe, D., Lochner, S., 2008. Erdgas für Europa: Die EWIGAS 2008 Prognose. *Zeitschrift für Energiewirtschaft* 32(1): 22–29.

Boussena, S., Locatelli, C., 2013. Energy institutional and organizational changes in EU and Russia: revisiting gas relations. *Energy Policy* 55: 180–189.

Buzan, B., Waever, O., de Wilde J., 1998. *Security: A New Framework for Analysis*. Boulder, CO: Lynne Rienner Publishers,

Cafruny, A., Rosenthal, G., (eds) 1993. *The State of the European Community*. New York: Lynne Rienner.

Chester, L., 2010. Conceptualizing energy security and making explicit its polysemic nature. *Energy Policy* 38(2): 887–895.

Chevalier, J.-M., 2006. Security of energy supply for the European Union. European Review of *Energy Markets* 1(3): 1–20.

Ciutâ, F., 2010. Conceptual notes on energy security: total or banal security? *Security Dialogue* 41(2): 123–144.

Clawson, P., 1998. Is energy security a meaningful concept? *Defense Journal* special report. See www.defencejournal.com/march98/energysecurity.htm.

Coase, R., 1998. The new institutional economics. *American Economic Review* 88(2) (papers and Proceedings of the Hundred and Tenth Annual Meeting of the American Economic Association, May): 72–74.

Cohen, G., Joutz, F., Loungani, P., 2011. Measuring energy security: trends in the diversification of oil and natural gas supplies. *Energy Policy* 39(9): 4860–4869.

Constantini, V., Gravecca, F., Markandya, A., Vicini, G., 2007. Security of energy supply: comparing scenarios from a European perspective. *Energy Policy* 35(1): 210–226.

Corbey, D., 1995. Dialectical functionalism: stagnation as a booster of European integration. *International Organization* 49(2): 253–284.

Correljé, A. F., Linde van der, J. G., 2006. Energy supply security and geopolitics: a European perspective. *Energy Policy* 34(5): 532–543.

De Bruijn T., 2002. Transforming regulatory systems: multilevel governance in a European context. See http://doc.utwente.nl/48277/1/Paper_Berlin_2002_De_Bruijn_revised.pdf.

Dehousse, F., 2008. *The Coming Energy Crash and Its Impact On the European Union*. Ghent: Academia Press.

Delaquil, P., Goldstein, G., Nelson, H., Peterson, T., Roe, S., Rose, A., Wei, D., Wennberg, J., 2012. *Developing and Assessing Economic, Energy, and Climate Security and Investment Options for the US*. 2012 International Energy Workshop Paper. Washington, DC: Center for Climate Strategies.

Demski, C., Poortinga, W., Pidgeon, N., 2014. Exploring public perceptions of energy security risks in the UK. *Energy Policy* 66: 369–378.

De Wilde, J. H., 1991. *Saved From Oblivion: Interdependence Theory in the First Half of the 20th Century*. Boston, MA: Dartmouth Publishing,

Dickel, R., Hassanzadeh, E., Henderson, J., Honoré, A., El-Katiri, L., Pirani, S., Rogers, H., Stern, J., Yafimava, K., 2014. *Reducing European Dependence on Russian Gas: Distinguishing Natural Gas Security from Geopolitics*. OIES paper NG92. Oxford: Oxford Institute for Energy Studies.

Diez, T., Wiener, A., 2003. *European Integration Theory*. Oxford: Oxford University Press.

European Commission, 2011. *Proposal for a Regulation on Guidelines for Trans-European Energy Infrastructure and Repealing Decision Number 1364/2006/EC*. COM(2011) 658 final. Brussels: European Commission. See http://eur-lex.europa.eu/LexUriServ/LexUriServ.do?uri=COM:2011:0658:FIN:EN:PDF.

European Commission, 2012. *Communication: Making the Internal Energy Market Work*. COM(2012) 663 final. Brussels: European Commission.

Gasmi, F., Oviedo, J. D., 2010. Investment in transport infrastructure, regulation, and gas-to-gas competition. *Energy Economics* 32(3): 726–736.

Goldman, M. I., 2008. *Petrostate: Putin, Power and the New Russia*. Oxford: Oxford University Press.

Goldthau, A., 2008. Russia's energy weapon is a fiction (with commentary). *Europe's World* (8): 36–41.

Helm, D., 2002. Energy policy: security of supply, sustainability and competition. *Energy Policy* 30(3): 173–184.

Helm, D., 2014. The European framework for energy and climate policies. *Energy Policy* 64: 29–35.

Henderson, J., Pirani, S., (eds) 2014. *The Russian Gas Matrix: How Markets Are Driving Change*. Oxford: Oxford Institute for Energy Studies.

Hirschl, B., 2009. International renewable energy policy: between marginalization and initial approaches. *Energy Policy* 37(11): 4407–4416.

Högselius, P., 2012. *Red Gas: Russia and the Origins of European Energy Dependence*. New York: Palgrave Macmillan.

Hooghe, L., Marks, G., 2001. *Multi-Level Governance and European Integration*. Lanham, MD: Rowman & Littlefield.

Hooghe, L., Marks, G., 2003. Unraveling the central state, but how? Types of multi-level governance. *American Political Science Review* 97(2): 233–243.

Hughes, L., 2009. The four R's of energy security. *Energy Policy* 37(6): 2495–2461.

International Energy Agency, 2001. *Toward a Sustainable Energy Future*. Paris: OECD/IEA.

Jamasb, T., Pollitt, M., 2008. Security of supply and regulation of energy networks. *Energy Policy* 36(12): 4584–4589.

Johnson, C., Boersma, T., 2013. Energy (in)security in Poland, the case of shale gas. *Energy Policy* 53: 389–399.

Kalicki, J. H., Goldwyn, D. L., (eds) 2005. *Energy and Security: Toward a New Foreign Policy Strategy*. Washington, DC: Woodrow Wilson Centre Press.

Klare, M. T., 2001. *Resource Wars: The New Landscape of Global Conflict*. New York: Henry Holt & Company.

Kratochvíl, P., Tichy, L., 2013. EU and Russian discourse on energy relations.

Energy Policy 56: 391–406.

Kruyt, B., Van Vuuren, D. P., De Vries, H. J. M., Groenenberg, H., 2009. Indicators for energy security. *Energy Policy* 37(6): 2166–2181.

Le Coq, C., Paltseva, E., 2012. Assessing gas transit risks: Russia vs. the EU. *Energy Policy* 42: 642–650.

Lipson, C., Cohen, B., (eds) 1999. *Theory and Structure in International Political Economy*. Cambridge, MA: MIT Press.

Manley, D. K., Hines, V. A., Jordan, M. W., Stoltz, R. E., 2013. A survey of energy policy priorities in the United States: energy supply security, economics, and the environment. *Energy Policy* 60: 687–696.

Matláry, J. H., 1993. Beyond intergovernmentalism: the quest for a comprehensive framework for the study of integration. *Cooperation and Conflict* 28(2): 181–208.

McGowan, F., 2011. Putting energy insecurity into historical context: European responses to the energy crises of the 1970s and 2000s. *Geopolitics* 16: 486–511.

Meyer, N. I., 2003. Distributed generation and the problematic deregulation of energy markets in Europe. *International Journal of Sustainable Energy* 23(4): 217–221.

Milina, V., 2007. Energy security and geopolitics. *Connections* 6(4): 25–44.

Mitrany, D., 1965. The prospect of integration: federal or functional. *Journal of Common Market Studies* 4(2): 119–149.

Mitrany, D., 1975. *The Functional Theory of Politics*. New York: St. Martin's Press.

Monaghan, A., 2007. *Russia and the Security of Europe's Energy Supplies: Security in Diversity?* Special series 07/02. Shrivenham: Conflict Studies Research Centre, Defence Academy of the United Kingdom.

Monni, S., Raes, F., 2008. Multilevel climate policy: the case of the European Union, Finland and Helsinki. *Environmental Science and Policy* 11: 743–755.

Moravcsik, A., 1994. *Why the European Union Strengthens the State: Domestic Politics and International Cooperation*. CES Working Paper (52). Cambridge, MA: Center for European Studies, Harvard University. See http://aei.pitt.edu/9151/1/Moravcsik52.pdf.

North, D. C., 1991. Institutions. *Journal of Economic Perspectives* 5(1): 97–112.

Nuttall, W. J., Manz, D. L., 2008. A new energy security paradigm for the twenty-first century. *Technological Forecasting and Social Change* 75: 1247–1259.

Oettinger, G., 2010. Lighting it up: securing the future of Europe's energy supply. *European View* 9: 47–51.

Özcan, S., 2013. Securitization of energy through the lenses of Copenhagen School. *West East Journal of Social Science* 2(2): 57–72.

Papadopoulos, Y., 2007. Problems of democratic accountability in network and multilevel governance. *European Law Journal* 13(4): 469–486.

Pickl, M., Wirl, F., 2010. Enhancing the EU's energy supply security: an evaluation of the Nabucco project and an introduction to its open season capacity allocation process. *Zeitschrift für Energiewirtschaft* 34: 153–161.

Pirani, S., Stern, J., Yafimava, K., 2009. *The Russo-Ukrainian Gas Dispute of January 2009: A Comprehensive Assessment*. Oxford: Oxford Institute for Energy Studies.

Pointvogl, A., 2009. Perceptions, realities, concession: what is driving the integration of European energy policies? *Energy Policy* 37(12): 5704–5716.

Proedrou, F., 2012. *EU Energy Security in the Gas Sector: Evolving Dynamics, Policy Dilemmas, and Prospects*. Farnham: Ashgate Publishing.

Rabe, B. G., 2007. Beyond Kyoto: climate change policy in multilevel governance systems. *Governance: An International Journal of Policy, Administration and Institutions* 20(3): 423–444.

Rosamond, B., 2000. *Theories of European Integration.* New York: Palgrave Macmillan.

Rosenau, J. N., (ed.) 1969. *Linkage Politics: Essays on the Convergence of National and International Systems.* New York: Free Press.

Rosenau, J. N. 2004. Strong demand, huge supply: governance in an emerging epoch. In I. Bache, M. Flinders (eds), *Multi-level Governance,* 31–48. Oxford: Oxford University Press.

Roth, M., 2011. Poland as a policy entrepreneur in European external energy policy: towards greater energy solidarity vis-à-vis Russia? *Geopolitics* 16(3): 600–625.

Scheepers, M., Seebregts, A., De Jong, J., Maters, H., 2006. *EU Standards for Energy Security of Supply.* ECN-C-06-039/CIEP. The Hague: ECN Clingendael International Energy Programme.

Schmidt-Felzmann, A., 2011. EU member states' energy relations with Russia: conflicting approaches to securing natural gas supplies. *Geopolitics* 16: 574–599.

Schmitter, P. C., 1969. Three neo-functional hypotheses about international integration. *International Organization* 23(1): 161–166.

Smith, K. C. 2006. Security implications of Russian energy policies. See www.ceps.eu/book/security-implications-russian-energy-policies.

Smith Stegen, K., 2011. Deconstructing the "energy weapon": Russia's threat to Europe as case study. *Energy Policy* 39: 6505–6513.

Söderbergh, B., Jakobsson, K, Aleklett, K., 2010. European energy security: an analysis of future Russian natural gas production and exports. *Energy Policy* 38(12): 7827–7843.

Stec, S., Baraj, B., (eds) 2009. *Energy and Environmental Challenges to Security: Proceedings of the NATO Advanced Research Workshop on Energy and Environmental Challenges to Security, Budapest Hungary, November 2007.* Dordrecht: Springer Science and Business Media.

Stern, J., 2009. *Future Gas Production in Russia: Is the Concern about Lack of Investment Justified?* Oxford: Oxford Institute for Energy Studies.

Talus, K., 2013 *EU Energy Law and Policy: A Critical Account.* Oxford: Oxford University Press.

Toft, P., Duero, A., Bieliauskas, A., 2010. Terrorist targeting and energy security. *Energy Policy* 38(8): 4411–4421.

Tranholm-Mikkelsen, J., 1991. Neo-functionalism: obstinate or obsolete? A reappraisal in the light of the new dynamism of the EC. *Journal of International Studies* 20(1): 1–22.

Trombetta, J., 2012. *European Energy Security Discourses and the Development of a Common Energy Policy.* Working paper no 2. Groningen: Energy Delta Gas Research.

Umbach, F., 2010. Global energy security and the implications for the EU. *Energy Policy* 38(3): 1229–1240.

Van der Linde J. G., Amineh, M. Correljé, A. F., De Jong, D., 2004. *Study on Energy Supply Security and Geopolitics.* Report prepared for DG TREN, contract number TREN/C1-06-2002, ETPA programme, January. The Hague: Clingendael International Energy Programme.

Von Hirschhausen C., 2008. Infrastructure, regulation, investment and security of supply: a case study of the restructured US natural gas market. *Utilities Policy* 16(1): 1–10.

Williamson, O. E., 1998. The institutions of governance. *American Economic Review* 88(2) (papers and Proceedings of the Hundred and Tenth Annual Meeting of the American Economic Association, May): 75–79.

Williamson, O. E., 2000. The new institutional economics: taking stock, looking ahead. *Journal of Economic Literature* 38: 595–613.

Yergin, D., 2006. Ensuring energy security. *Foreign Affairs* 85(2): 69–82.

Part II

Energy policy in the European Union

3 Status quo in European Union energy policy

Introduction

Numerous policy makers have contributed to attempts to create a single EU energy policy so far. And it seems no hyperbole to suggest that numerous will follow. As elaborated on in the first two chapters of this book, most focus of both academic contributions and policy makers has been on developing one single EU energy market (liberalization). The emphasis on unbundling issues and external relations with major suppliers within the triplet of gas directives that has been published so far underlines that statement, as is demonstrated in more detail throughout this chapter. In addition the EC itself stated in one of its considerations regarding Decision 1229/2003/EC that new priorities in energy infrastructure "stem from the creation of an open and more competitive internal energy market, as a result of the implementation of Directive 96/92/EC ... and of Directive 98/30/EC ... concerning common rules for the internal market in natural gas."[1]

Since the publication of the first Electricity Directive by the EC in 1996, that same EC has shown enthusiasm to develop trans-European energy networks.[2] Yet where the EC has been relatively effective in creating a more European energy market, infrastructure and also regulation has predominantly remained the domain of the member states, and transboundary infrastructure the domain of private and semi-public market players. Since 2011 renewed steps of the EC on both these elements of the energy system can be identified. The achievements during the past two decades, in particular with regard to energy infrastructure, are not insignificant. As this overview shows, in light of the Ukraine crisis it has now become "bon ton" to argue for increased market integration as a means to harness against market abuse of dominant suppliers, as the fear for supply disruptions has been revitalized. As is shown throughout this book however, the goals set by the EC to complete the internal energy market by 2014 were never realistic, and in a business as usual scenario it will take member states at least until the end of the decade to reach this target. Moreover, there are some European member states (i.e. Poland and Lithuania) that appear to have lost their confidence that further European integration is going to deliver energy security. Instead, these countries have chosen to implement regulatory regimes that favor purchasing more expensive LNG in detriment of Russian natural gas. As is argued later in this book, these decisions should probably be seen as expensive insurance policies against market power abuse. Whether markets will eventually dictate that Russian gas is in fact backed out of these countries, remains to be seen.

This chapter successively describes the status quo in terms of legislation and actual steps taken so far by the EC and the member states on the gas market, gas infrastructure and regulatory authorities and discusses these observations and their consequences in terms of decision-making structures. The chapter is based predominantly on primary sources (i.e. existing legislation and regulations, interviews with experts and policy makers, and academic contributions, all for the period up to November 2014).

Markets

Rules and regulation

Since the early 1990s the gas market has been subject to efforts of the EC to advance the internal energy system. Its main arguments notably were "security of supply and environment protection." The first directives focused on the facilitation of non-discriminatory transit of natural gas between high-pressure transmission grids and the improvement of the transparency of gas and electricity prices charged to industrial end-users.[3]

Several years later the EC estimated that the time was right to further develop the internal market for natural gas.[4] It had taken its time, as for instance underlined by this consideration: "whereas the internal market in natural gas needs to be established gradually, in order to enable the industry to adjust in a flexible and ordered manner to its new environment and in order to take account of the different market structures in the Member States."[5] The directive laid down common rules for the transmission, distribution, supply and storage of natural gas. Member states were expected to guarantee non-discriminatory access to their market for undertakings to invest in gas facilities and to make available minimum technical standards regarding storage facilities or distribution and transmission systems. Moreover integrated natural gas companies had to keep separate accounts for gas and non-gas activities, in order to prevent cross-subsidization and distortion of competition.[6] Finally, the directive contained elements that strived to arrange third-party access (TPA), gradual market opening within the member states and reduced market domination of incumbent players.

In 2003 the EC established "significant shortcomings" in the desired completion of the market and launched the second Gas Directive.[7] One of the important elements was the call for transmission system operators that were part of vertically integrated undertakings to be "at least independent in terms of legal form, organization and decision making from other activities not relating to transmission." This implied the EC was taking a clearer stand on so-called unbundling of integrated energy companies. The argument was that these constructions instigated market distortions (in case of cross-subsidization) and at minimum did not bring about a higher level of transparency. The new directive also aimed to improve the access of new suppliers to the market and gave consumers the ability to switch freely between gas suppliers.[8] In addition the EC recognized the importance of an active role of consumers for the market to function properly. Hence a number of measures were recorded in the directive to ensure among others transparent contract conditions and dispute settlement mechanisms. Finally, the directive prompted member states to appoint system operators that have responsibility for among others safety, reliability and interconnection facilities and to appoint independent regulators to monitor transparency, discrimination, the level of competition and the tariffs used by system operators. As part of the Second Energy Package a network of European national regulatory authorities was established to safeguard the completion of the internal market, called ERGEG.[9] This collaboration is elaborated on in more detail later in this chapter.

Nonetheless, not long after the publication and (partial) implementation of the new legislation the EC concluded that these rules and measures did "not provide the necessary framework for achieving the objective of a well-functioning internal market."[10] One of the main arguments of the EC to propose new legislation was that contractual congestion of infrastructure still posed serious access and market integration problems, and that effective unbundling of network companies had not been carried out throughout the Community. The EC stated that "without effective separation ... there is a risk of discrimination not only in the operation of the network but also in the incentives for vertically integrated undertakings to invest adequately in their networks."[11]

Hence the EC launched the so-called Third Energy Package, containing three regulations and two directives, aiming to amplify consumer rights and putting more emphasis on the benefits of the internal market, facilitating cross-border trade, stimulating investments and finally improving the coordination of regulation on the EU level.[12] The two directives in the package mainly focused on broadening consumer rights, for example, by requiring member states to ensure that consumers can effectively switch between suppliers within three weeks – apart from contractual obligations – and to ensure that consumers have one single point of contact, that provides them with all the information needed to be well informed and actively participate in the market. Directive 73/2009, however, also dealt with the meanwhile awkward file of unbundling. It turned out that the EC had suffered too much opposition from in particular Germany and France. This was evidenced by article 9 of the directive, which offered the possibility of an alternative next to full ownership unbundling, although the latter had always been propagated by that same EC. This directive proposed two alternatives: the so-called Independent System Operator (ISO) and the "third way" of an Independent Transmission Operator (ITO). Under ISO member states had the opportunity to leave transmission networks under the ownership of energy companies, but operation and control of the day-to-day business came into the hands of an independent system operator. In practice an ITO meant that the Transmission System Operator (TSO) remained within the integrated company and the relevant assets of the TSO remained on the balance sheet. The directive did introduce regulatory arrangements in order to guarantee the independence of the ITO from the integrated company. The essential difference rested in the unchanged balance sheet of integrated companies. This was important for integrated companies, because transmission systems in general form a substantial part of that balance sheet. Integrated companies would have lost a significant amount of financial impact pressure if full ownership unbundling had been applied.

The three regulations in the Third Package put more emphasis on in particular the infrastructure and regulation part of the EU energy system. The EC for instance established an alliance of transmission companies, called European Network of Transmission System Operators for Gas (ENTSO-G) and European Network of Transmission System Operators for Electricity (ENTSO-E).[13] In addition a new European institution to improve the regulatory framework was founded, and called the Agency for the Cooperation of Energy Regulators (ACER).[14] These institutions are discussed later in this chapter. Regulation 715/2009 furthermore, again, focused on TPA and transparency requirements.

As is explored in more detail in Chapter 7, having laws and regulations does not necessarily mean that they are also implemented or acted upon. As of writing, more than a handful of member states are still working on the implementation of the aforementioned legislation.

Gas infrastructure

Rules and regulation

Since 1996 the EC has more actively included infrastructure in its considerations as part of guaranteeing EU energy security. The first Gas Directive could be labeled as rather straightforward and having potentially substantial consequences, for instance with regard to the earlier mentioned unbundling discussion. However, regarding gas infrastructure the EC for a long time seemed more reticent, likely because there was no agreement among member states on how to approach this issue.

Decision 96/391/EC paved the way for the EC to make co-investments in energy infrastructure in order to reduce energy supply costs and thus facilitate economic growth, to stimulate employment and enhance competitiveness, next to the existing financial instruments such as the Structural Funds, the European Investment Fund and support from the European Investment Bank. This financial angle was not entirely new, since Regulation 2236/95/EC already described procedures and conditions for granting Community aid to projects of common interest in the field of trans-European networks in, among others, energy infrastructure.[15] The decision also focused on promoting the technical cooperation between member states in order to enhance the realization of important trans-European infrastructural projects.[16] In addition the EC had also drawn up an overview of projects that deserved priority in this matter, a committee to support it in its work and laid down the commitment to report to the EP, the Council, the Economic and Social Committee and the Committee of the Regions.

These intentions were recorded in the first version of the guidelines for trans-European energy networks, or so-called TEN-E Guidelines (hereafter referred to as TEN-E).[17] These guidelines have been renewed twice since their first appearance, resulting in the most recent Decision 1364/2006/EC. The TEN-E listed and ranked energy infrastructure projects eligible for Community assistance. The objectives of TEN-E were to contribute to a more effective operation of the EU energy system – for which sufficient interconnection and interoperability are essential – and to guarantee security and diversification of supply (think, for example, of sufficient interconnection with countries at the frontiers of the EU). In addition TEN-E aimed to strengthen the territorial cohesion within the EU by reducing the isolation of remote regions. Finally, it aimed to promote sustainable development by improving the links between renewable energy production installations and using more efficient technologies.

Projects eligible for Community assistance are ranked in three categories:[18]

- Projects of common interest related to electricity and gas networks in Decision 1364/2006/EC that display potential economic viability. The latter was assessed by means of a cost-benefit analysis in terms of the environment, security of supply and territorial cohesion.
- Priority projects were selected from the category above, but had to have a significant impact on the functioning of the internal market, on the security of supply and/or the use of renewable energy sources. These projects had priority for granting financial assistance.
- Finally, projects of European interest were priority projects of either a cross-border nature or which had a significant impact on transmission capacity. They had priority for granting under the TEN-E budget and were given extra attention regarding their funding.

However, despite the good intentions, in general no large financial support schemes were available to finance infrastructural projects. The TEN-E budget consists of around €20 million per year and is mainly used to finance feasibility studies.[19] Rather, TEN-E provided a framework that stressed the importance of the proper facilitation and completion of these infrastructural projects by the European member states, in particular those of European interest. The proceedings were monitored by EU officials and in case of delays or serious difficulties a coordinator could step in to speed up the process.[20] However, the authority of the coordinator did not exceed facilitating the discussion between the institutions involved and reporting to the EC on the proceedings of the relevant project on a yearly basis.

In 2007 the EC reported to both the Council and the EP about the proceedings of the projects as listed under TEN-E adopted in 2006.[21] The communication on the Priority Interconnection Plan (PIP) gave an update about the progress of the 42 projects of European interest listed in TEN-E. Supposedly 60 percent of electricity projects were behind schedule due to complexity and lack of harmonization in planning and authorization procedures. The gas projects received slightly better reports, but in particular LNG-projects and interconnectors required special attention. PIP gave a detailed overview of the relevant policies in this matter, addressing TEN-E, the first legal framework to safeguard security of gas supply (i.e. Directive 2004/67/EC) and reiterated the support that the Council had given in 2006 to realize an interconnected, transparent and non-discriminatory internal energy system. The EC reported that there was insufficient progress, a lack of non-discriminatory network access, and lack of an equally effective level of regulatory supervision. Furthermore the EU lacked a common regulatory framework to coordinate investments in infrastructure and it also required mechanisms to properly coordinate technical standards, balancing rules and gas quality.[22] Most under-investments were due to insufficient unbundling throughout the member states, according to the EC. In particular cross-border infrastructure was neglected, leading to the conclusion that "with infrastructure investment as it currently stands, the EU will not be able to construct a real single market."

PIP focused specifically on the projects, labeled projects of European interest under TEN-E, which experienced significant delays. The analysis showed that the most urgent problems were found in the electricity sector, however, "risks for pipeline investments crossing multiple frontiers are perceived to be growing." The EC expressed her concern about the lack of a framework for investment. It argued that the "current market design does not create incentives for efficient transmission investment." With the budget of TEN-E being limited, the EC declared itself in favor of evaluating the current financial framework.

The financial crisis that started in 2007 in the US provided the EC with an excellent momentum to actively invest *itself* in key infrastructure throughout the EU.[23] Moreover, the gas supply disruption in 2009 once again underlined the importance to address energy security, according to the EC in, among others, her Second Strategic Energy Review.[24] Within its

European Economic Recovery Plan it reserved €1.75 billion for "key strategic interconnections."[25] However, with the financial and economic crisis unfolding, projects that had been planned were delayed or withdrawn. This urged both EP and Council to establish the European Energy Program for Recovery (EEPR) as a financial instrument to boost investment and stimulate the rapid realization of the EU energy and climate policy objectives.[26] The total envelope of the EEPR encompassed close to €4 billion, of which nearly €2.4 billion were allocated to a series of targeted gas and electricity infrastructure projects. The EC claimed the allocation of financial means to be successful and expected the EEPR grants to mobilize up to €22 billion of private sector investment within five years.

Box 3.1 European Union energy policy in the Lisbon Treaty

Following the adoption of the Lisbon Treaty in late 2007, some authors concluded that energy policy had become a shared competence between member states and the EU (e.g. Trombetta 2012). Yet it is worth examining what this shared competence entails. The following articles in the Lisbon Treaty refer to energy and energy policy:

- Article 122 ascribes the EU competences in case of energy supply disruptions (this used to be Article 100 TEC).
- Article 170 allows the EU to contribute to trans-European energy networks (this used to be Article 154 TEC).
- Article 192 gives the EU mandate to set environmental standards, also when the national energy mix is concerned (this used to be Article 175 TEC).
- The newly inserted Article 194 states that EU policy aims to ensure the functioning of the energy market, ensure security of supply, promote energy efficiency and renewable energy, and promote the interconnection of energy networks. However, this "shall not affect a member state's right to determine the conditions for exploiting its energy resources, its choice between different energy resources, and the general structure of its supply."

Source: Consolidated version of the Treaty on the European Union and the Treaty of the Functioning of the European Union, 2010/C, 83/01

Until then, this round of investments turned out to be an incident and the structural lack of an infrastructure policy started to show effect. The reports of the EC to the relevant European institutions on the implementation of the trans-European energy networks as laid down in TEN-E underline this statement.[27] The EC in this report evaluated the contributions of TEN-E to its initial goal (i.e. to guarantee security of supply). It concluded that projects labeled "of European interest" in general had successfully been addressed by the responsible coordinators. It was hardly surprising that the other two lists, regarding priority projects and projects of common interest, had not been dived into as ambitiously, if only for the huge amount of projects selected under TEN-E.[28]

The EC also connected TEN-E to the renewed European goals on energy security and the 20-20-20 targets on sustainability. It concluded that "the impact of TEN-E has been less relevant in dealing with the more recent challenges concerning the EU's strategic energy policy goals and targets."[29] Thus, the EC concluded that a new approach towards infrastructure was necessary in order to guarantee long-term energy security within the EU. With the support of the Council to thoroughly evaluate TEN-E in November 2010 the EC published its Blueprint for an integrated European energy network.[30] It stated that not only to enable the optimization of the internal market for energy, but also for reasons of security of supply, the integration of renewable energy, the increase of energy efficiency and to enable consumers to benefit from all these attainments, "a new EU energy infrastructure policy is needed to coordinate and optimize network development on a continental scale." The EC concluded that the Third Package had laid the basis for a European approach towards infrastructure planning and investment and a more European approach of regulation by relevant national authorities.[31]

Direct involvement of the EC in energy infrastructure affairs proved to be a delicate matter. In February 2011 the European Council concluded that streamlining and improving authorization procedures was important "while respecting national competences and procedures, for the building of new infrastructure." Later on the document reads as follows:

> the bulk of the important financing costs for infrastructure investments will have to be delivered by the market, with costs covered through tariffs … However, some projects that would be justified from a security of supply/solidarity perspective, but are unable to attract enough market-based finance, may require some limited public finance.[32]

Implicitly the Council expressed limited enthusiasm for EC involvement in energy infrastructure, but recognized that exceptions could be considered.[33] It therefore asked the EC to draft an overview with more detailed figures on the required investments, suggestions on how to deal with the financing of these projects, and other possible obstacles to realize these investments in energy infrastructure.

So the EC did in June 2011, sketching a rough estimate of energy infrastructure requirements worth €210 billion for the decade ahead, of which roughly €70 billion was reserved for gas transmission pipelines, storage, LNG/CNG terminals and reverse flow infrastructure. Other estimates are

higher, such as the Ten Year Network Development Plan published by
ENTSO-G in March 2011, which estimated the figure at €89 billion. A
study commissioned by Roland Berger concluded that investment volumes
for natural gas during the period up to 2020 would increase by 30 percent.[34]
These investments are at risk of not being delivered by 2020 due to obstacles related to permit granting and regulation and financing, as announced
in the impact assessment accompanying the 2010 infrastructure Blueprint.[35]
The EC stated that for certain energy infrastructure projects the current
framework was just insufficient, or "focus of national tariff setting frameworks on national networks and consumers as well as the pressure to keep
grid tariffs as low as possible in a context of low acceptability of structurally rising energy prices does not incentivize to invest in these projects."[36]
Characteristics of these projects could be higher regional than national
benefits, high technological risks, externalities such as an increase of
regional energy security, increased market competition, etc. Approximately
€60 billion of the projects would be subject to these obstacles.

To address the obstacles as described the EC in October 2011 published
its proposal for a Regulation on guidelines for trans-European energy infrastructure.[37] It intended to replace the TEN-E guidelines structure, by
recognizing that "the policy lacks focus, flexibility and a top-down
approach to fill identified infrastructure gaps." The EC concluded that the
main identified obstacles, problems related to permit granting, regulation
and financing, had to be addressed by formulating projects of common
interest that would:

- get a special permit granting procedure of maximum three years intro-
  ducing a so-called one-stop-shop in each member state;
- suggest an *ex ante* cross-border cost allocation mechanism and incentives commensurate with the risks incurred by the operator; and
- make them eligible for EU funding.[38]

Regarding eligibility of EU funding, the EC aimed to allocate €9.1 billion
for energy infrastructure out of a total budget of €50 billion for the period
of 2014–2020, proposed in the Regulation for a Connecting Europe
Facility.[39] As of early 2013 the proposed regulation had not been accepted
by the member states, despite repeated calls of the EC.[40] In July 2013 the
member states did reach an agreement, though the available budget for
energy infrastructure (both electricity and natural gas) had been watered
down to €5.1 billion.[41] The guidelines for financial eligibility had been
agreed on earlier in the spring of 2013, with the adoption of Regulation
347/2013. The Regulation established a list of 12 regional groups, which all
should adopt a proposed list of projects of common interest. All these projects by law have to fulfill at least one of the following criteria: market
integration, security of supply, competition, sustainability.[42] If these projects
fulfill the criteria, then some of them will be eligible for financial assistance

in the form of grants for studies and financial instruments, whereas others, depending on another set of criteria, may be eligible for financial assistance to physically construct infrastructure. It is important to keep in mind that the total funds for these infrastructure projects under CEF collectively cannot surpass €5.1 billion in the period up to 2020. Moreover, it is important to keep in mind that, when examining the energy infrastructure corridors in annex 1 of the Regulation, it is evident that all member states have been set up to receive a piece of this pie: though 4 priority gas corridors have been established, the member states that fall under them comprise all 28 (even though Croatia at the time of writing of the Regulation had not joined the EU) member states of the EU.

In October 2013 a first list of no less than 248 projects of common interest was adopted by the EC, and added as Annex 7 to the existing Regulation 347/2013.[43] In May 2014 the EC then released the first €750 million of financial support for infrastructure projects, for which – for reasons that have not been communicated – the total budget until 2020 now comprised €5.85 billion.[44] In order to qualify for financial support, projects have to be required to have a cross-border cost-allocation decision issued by the competent national regulatory authorities, or, in case they disagree, by the Agency for the Cooperation of Energy Regulators.[45] In October 2014 the EC released its first decision of co-financing of energy infrastructure based on its structural mandate under CEF.[46] Member states agreed to allocate €647 million to 34 projects, of which 16 were natural gas related. Of the 34 grants, 28 were feasibility studies, and 6 grants were allocated for construction projects (worth €555.9 million in total). According to the indicative list of projects selected for receiving financial assistance under CEF, the natural gas related projects that received funds for physical construction of infrastructure are an interconnector in Scotland (€33.7 million), a gas transmission pipeline in Lithuania (€27.5 million), and an interconnector between Poland and Lithuania with supporting infrastructure (€295.3 million).[47]

This overview of gas infrastructure has given a detailed indication of the difficulty to reach consensus within the EU when it comes to natural gas infrastructure, and particularly co-financing projects, all of which are in essence believed to be member state affairs. It is important to establish that there is a significant discrepancy between the estimated amounts of investments required (over €200 billion in the period up to 2020) and the level of agreement between the 28 member states to address the issues at hand (despite investments not being made, only marginal investments are being done in the coming years, and there is no reason to expect this is changing before 2020. This is even more remarkable in the context of what is often described as an eminent danger, namely (in parts of Europe) a dominant external gas supplier (Gazprom). This chapter now first continues with coordination and representation in relation to natural gas infrastructure.

Coordination and representation

A part of the new approach towards infrastructure had been the establishment of ENTSO-G.[48] In order to deal with the rather fragmentary nature of infrastructural companies throughout the EU, the EC established this cooperation in order to "promote the completion and functioning of the internal market in natural gas and cross-border trade and to ensure the optimal management, coordinated cooperation and sound technical evolution of the natural gas transmission network."[49] ENTSO-G focused in particular on cross-border investments and interoperability. Its tasks are here somewhat irreverently described as coordinating (i.e. formulating common network operation tools) and advising (i.e. drafting a non-binding Community-wide ten-year network development plan or formulate recommendations regarding technical cooperation between Community and third countries transmission system operators).[50] It is emphasized in the regulation that the network codes developed by ENTSO-G "are not intended to replace the necessary national network codes for non-cross border issues." However, one can argue that the establishment of ENTSO-G provided the EC for the first time with a more European view on access to infrastructure and making investments in infrastructure throughout the EU.

An increase in policy initiatives on the Community level was expected to catalyze an increase in representative forces on that level as well. Within the gas infrastructure sector this has happened with the establishment of Gas Infrastructure Europe (GIE), a non-profit representative organization towards all relevant European institutions. Its list of members gives another impression of the fragmented overview of infrastructural companies within the EU: 66 members from 26 countries, gathering transmission system operators, storage system operators and LNG terminal operators. Comparable to ENTSO-G, GIE focuses on the promotion of interoperability and the enhancement of cross-border activity in the EU gas system.

Next to GIE a number of representative organizations participate on the EU level, such as Eurogas (consisting of industry executives and specialists from the member states), the Technical Association of the European Natural Gas Industry (dealing with technical regulations and standards) and the European Research Group (group of companies with a strong base in R&D).

Regulatory authorities

Somewhat later than infrastructure, the EC at the end of 2003 established an independent advisory group on electricity and gas, or European Regulators Group for Electricity and Gas (ERGEG).[51] Its main task is to advise and assist the EC with the completion of the internal energy system, in particular when new regulation is concerned. The heads of the national regulatory authorities from the member states form the members of the organization.[52] Next to consultation and cooperation with the EC, ERGEG also sets out to facilitate these activities among the regulatory bodies in the member states.

The establishment of ERGEG, however, does not mark the first cooperation between European energy regulators. From 2000 onward ten regulatory authorities had been working together in order to exchange information and experience in relation to their common interest: the promotion of the internal energy market. In 2003 this cooperation resulted in the establishment of the Council of European Energy Regulators (CEER). Basically CEER pursues the same goals as ERGEG and the two institutions work closely together. Whereas CEER is concerned with the preparation of the work of ERGEG, the main difference rests in their respective voluntary nature versus the establishment by law.[53]

From 2003 onward the EC has been evaluating the achievements of the regulatory organizations working together more closely. In 2007 establishing a more solid regulatory framework was recognized as one of the key measures in order to complete the internal energy market, since the existing mechanisms were evaluated as inadequate to harmonize in particular technical standards and hence promote cross-border trade within the Union.[54] One of the options the EC proposed was to set up a new body at Community level to deal with this lacuna. In March 2009 the Council agreed to this option that resulted in the establishment of ACER.[55] Its most distinct tasks were the monitoring of the cooperation of transmission system operators within ENTSO-G (and its brother for electricity ENTSO-E), monitoring the progress of projects that particularly deal with interconnector capacity and providing a framework within which national regulatory authorities can cooperate. In addition ACER received a shared decision-making task regarding terms and conditions for access when cross-border infrastructure is concerned, in case the national regulatory authorities have not reached an agreement within six months or when these authorities request ACER to mediate.[56] In August of 2014 ACER reached the first decision of this kind, in the case of the Polish – Lithuanian interconnector that was mentioned before.[57] Lavrijssen-Heijmans and Hancher (in Arts *et al.* 2008) observed that alike ERGEG, ACER still was a hybrid organization, consisting of representatives of national regulatory authorities, and pointed at problems with political and legal accountability of this type of organizations. Despite the lack of an explicit basis for these organizations to formulate European policy, their activities can have far reaching effects. The lack of formal powers of ACER is confirmed by Coen and Thatcher (2008), who also pointed at its lack of resources and the absence of the right of initiative. They examined what motivated the EC, national governments and regulatory authorities to create these organizations. They concluded that the main argument lies in problems of coordination, in combination with the inability to agree on establishing a European regulatory authority, making this option "second best" (*ibid.*: 67). In September

2014 ACER indicated in a paper on its vision on the 2025 European energy market and its role herein that "adequate powers" should be given to the institution to effectively fulfill the monitoring responsibilities assigned to it.[58] New legislation would be required to expand its mandate.

Discussion

This chapter demonstrated the persistent attempts that the EC carried out in order to get a better understanding of the EU gas system and possibly firmer hold of EU energy policy. Zooming in on the different elements of the energy system it is evident that the EC's focus has been on the market since the early 1990s. The legislative history shows that the revolving element in these discussions has been unbundling of integrated energy companies. Combining the desires as expressed by the EC and the status quo regarding unbundling, the conclusion seems justified that the EC has backed down from this particular element of the energy system, for it settled for the compromise of ITO. However, the EC has countered this imperfection in the internal energy market by repealing the existing directive on security of supply and replacing it with a regulation on the same topic (this is discussed in detail in the next chapter).

Moving to infrastructure and regulatory authorities, the EC has made proposals for more European coordination and cooperation in this field since the late 1990s. Until recently the proposals were rather reticent and therefore not controversial, in contrast to the unbundling discussion. However, by now the EC is more on top of these elements of the energy system, due to the still existing problem of contractual congestion and the lagging investments in interconnections and challenges such as integrating expected amounts of sustainable energy into the energy system. In addition, the EC has earmarked infrastructure as a key to safeguard security of supply. By now it has become "bon ton" to state that the disruptions to gas supply in the 2000s could have been countered if the system had functioned better (European Commission 2010). As this analysis has also shown, that establishment has so far not resulted in more urgency to develop policies to help co-invest in infrastructure, or find other ways of attracting investments in gas infrastructure (which are discussed in Chapter 5). Finally, regulation continues to be mostly a national affair that is monitored and on occasion coordinated on the European level.

The EC's attention in terms of the EU energy system has initially been on the market element for over two decades and its attention has only recently been diversified towards other crucial elements of the energy system as such – namely infrastructure and regulation. That confirms the mismatch that was suggested in the introduction of this book (and is reflected in the academic debate about energy security as well). Furthermore, this chapter has demonstrated that the process of shaping the EU gas system's basic institutional framework is ongoing. It is generally the case that it is this institutional framework that serves as an incentive structure that creates opportunities for organizations to evolve (see North 1991: 109). Therefore, and given earlier observations that without massive discontent a sharp break from existing procedures should not be expected (Williamson 2000), it is likely that the EC has a long and expectedly bumpy road ahead to complete its internal energy system. Or, as Makholm (2012: 174) concluded: "the EU is in for some decades of work on the institutions that govern the way its pipelines are regulated, transact with customers, facilitate or impede the growth of a competitive gas market, and promote the security of its gas supplies." Policy makers in the coming years will demonstrate whether the crisis in Ukraine is such a moment of massive discontent that spurs a sharp break from existing procedures, though at the time of writing there were, despite all the rhetoric and upheaval, not many signs that would support that notion.

So why have relations between EU institutions and member states evolved in the way they have? It is an easy question, but unfortunately one without easy answers. The observed initial focus of the EC on the energy markets seems to fit the timeframe of the 1990s and early 2000s, in which liberalization and privatization of markets were fashionable. Other crucial elements of the energy system may have been less appealing to Brussels' civil servants, and it is not unlikely that there was just not enough manpower to do everything at once. Probably most importantly, EC policy makers at the time may have had the idea that there was just no incentive at the member state level to collaborate on these matters: after all, infrastructure investments and designing of regulatory regimes were national affairs, and they are to date. It is worth noting that throughout the process hesitance to shift power to the supranational level can be observed, for instance with regard to the unbundling of integrated energy companies or the mandate for EU institutions to structurally finance energy infrastructure. Also, the meager mandate and limited financial means of ACER suggest a compromise rather than a full-fledged choice for a supranational approach. However, the motives behind these examples may be different. Whereas hesitance to unbundle integrated energy companies has been observed in countries with histories of industrial policy (most notably Germany and France), it may perchance be explained as protectionist behavior of national entities. In contrast, more liberally oriented member states such as Great Britain and the Netherlands were more energetic in implementing this particular legislation and unbundled their integrated energy companies. With regard to the financing of energy infrastructure, member states have so far predominantly stressed that this is a national affair. In the rare case where European institutions did receive a mandate, it seems to be most important for the member states, given the politics involved in the allocation process, to get a piece of the pie. As a result, the available money in those rare instances did not necessarily flow to the places in the EU where it was most needed. An example of this was the allocation of funds from the European Economic

Recovery Plan to all member states, while arguably some parts of the EU energy system were more in need of those funds than others.[59] Another example is the allocation in October 2014, in the aftermath of the crisis in Ukraine, of financial means to construct an interconnector in Scotland. With all due respect, this is not necessarily a place where one would, again in the context of all the rhetoric about dependence on Russia, expect the EC to address energy security. Finally, with regard to regulation, here too member states have been reluctant to transfer authority to the supranational level. As this chapter has shown, ACER and its mandate seem to be a compromise between member states that are reluctant to transfer power on the one hand, and European institutions that aim to complete the internal energy market and claim to need power to do so on the other. Overall these dynamics suggest a struggle between European institutions and member states that is not likely to end any time soon.

Notes

1   Directive 1229/2003/EC consideration 2.
2   Council Decision 96/391/EC laying down a series of measures aimed at creating a more favorable context for the development of trans-European networks in the energy sector.
3   See Directive 91/296/EEC and Directive 90/377/EEC respectively.
4   Directive 98/30/EC. Take note that the first Electricity Directive was published two years earlier.
5   *Ibid.*
6   *Ibid.*, art. 13 sub 3.
7   Directive 2003/55/EC.
8   Note that from July 2004 industrial clients have had this privilege, whereas domestic consumers were treated equally starting July 2007. This mismatch gave suppliers the opportunity to adapt to the new circumstances.
9   Decision 2003/796/EC.
10 Directive 2009/73/EC, sub 5.
11 *Ibid.*, sub 6.
12 The relevant legislative documents in the package are Regulation 713–715/2009 and Directive 72–73/2009.
13 Regulation 715/2009, art. 8.
14 Regulation 713/2009 establishing an Agency for the Cooperation of Energy Regulators.
15 Regulation 2236/95/EC art. 1, following EC Treaty art. 155 sub 1.
16 Decision 96/391/EC art. 2 sub 1.
17 Decision 1254/96/EC, later repealed by Decision 1229/2003/EC and then by 1364/2006/EC.
18 Decision 1364/2006/EC, art. 6–8.
19 *Ibid.*, consideration 17. See also Regulation (EC) 680/2007 for the TEN Financial Regulation. Note in addition that the European Investment Bank has played a more important role in financing TEN-E projects.
20 However, no coordinator can be appointed without agreement of the member states involved and not after the EP is consulted, Decision 1364/2006/EC art. 10 sub 1.
21 COM(2006) 846 final/2, February 23, 3007.

22 *Ibid.*
23 Note that the EC did need a mandate from the Council to draft its Economic Program for Recovery, which it got in December 2008.
24 COM(2008) 781 final, November 13, 2008.
25 IP/09/142, January 28, 2009. The rest of the total budget of €3.5 billion for energy projects was intended for carbon capture and storage and offshore wind projects.
26 Regulation (EC) 663/2009.
27 SEC(2010)505 final, COM(2010)203 final, May 4, 2010.
28 To illustrate this, the priority projects list contains around 140 electricity projects and 100 gas projects, whereas projects of common interest count around 160 and 120 respectively.
29 COM(2010) 203 final, p. 7, http://ec.europa.eu/transparency/regdoc/rep/1/2010/EN/1-2010-203-EN-F1-1.Pdf.
30 COM(2010) 677/4.
31 *Ibid.*, para 2.6, p. 8.
32 Conclusions of the European Council (February 4, 2011), EUCO 2/1/11 REV 1 published March 8, 2011.
33 This is based on for instance the position as expressed by then Dutch Minister of Economic Affairs Mr. Verhagen in Parliamentary debates in early 2011.
34 SEC(2011) 755 final, page 4.
35 SEC(2010) 1395.
36 SEC(2011) 755 final, p. 6.
37 COM(2011) 658 final, 2011/0300 (COD).
38 2011/0300 (COD), p. 6ff..
39 COM(2011) 665, 2011/0302 (COD).
40 See for instance COM(2012) 663 final, Making the internal energy market work, p. 16.
41 Press release 11439/13, PRESSE 285, July 10, 2013.
42 Regulation 347/2013, article 4.
43 Commission delegated Regulation 1391/2013.
44 EC press release IP/14/547.
45 In August 2014 ACER took its first and to date only decision about cross-border cost allocation, as the national regulators of Poland and the Baltic states could not agree on the allocation of costs for the interconnector between Poland and Lithuania, in which Poland is the cost bearer and the Baltic states the net benefiting countries. See www.acer.europa.eu/Media/News/Pages/ACER-adopts-a-decision-on-the-allocation-of-costs-for-the-Gas-Interconnection-project-between-Poland-and-Lithuania.aspx.
46 EC press release IP/14/1204.
47 See indicative list published on October 29, 2014 at http://ec.europa.eu/energy/infrastructure/pci/pci_en.htm.
48 Regulation (EC) 715/2009.
49 Regulation (EC) 715/2009, art. 4.
50 For a broader description of the tasks of ENTSOG, see Regulation (EC) 715/2009, art. 8, sub 3.
51 Decision 2003/796/EC.
52 Earlier in 2003, member states were urged to address one or more competent bodies with the function of regulatory authorities, basically to monitor the implementation of existing and new legislation, as arranged in Directives 2003/54/EC and 2003/55/EC.
53 CEER is officially a non-profit organization under Belgian law.
54 COM(2007) 1 final, 10.1.2007.

55 Regulation (EC) 713/2009.
56 *Ibid.*, art. 8 sub 1. Terms and conditions for access to cross-border infrastructure include a procedure for capacity allocation, a time frame for allocation, shared congestion revenues and levying of charges on the users of the infrastructure.
57 See www.acer.europa.eu/Media/News/Pages/ACER-adopts-a-decision-on-the-allocation-of-costs-for-the-Gas-Interconnection-project-between-Poland-and-Lithuania.aspx.
58 Energy Regulation: A Bridge to 2025 Conclusions Paper – Recommendation of the Agency on the regulatory response to the future challenges emerging from developments in the internal market.
59 For an overview of the projects under the EERP, see http://europa.eu/rapid/press-release_IP-09-142_en.htm.

References

Arts, G., Dicke, W., Hancher, L., (eds) 2008. *New Perspectives on Investments in Infrastructures*. Amsterdam: Amsterdam University Press.
Coen, D., Thatcher, M., 2008. Network governance and multi-level delegation: European networks of regulatory agencies. *Journal of Public Policy* 28(1): 49–71.
Council Regulation 2236/95 of September 18, 1995 laying down general rules for the granting of Community financial aid in the field of trans-European networks.
Council Regulation 347/2013 of April 17, 2013 on guidelines for trans-European energy infrastructure.
Decision 96/391/EC of March 28, 1996 laying down a series of measures aimed at creating a more favourable context for the development of trans-European networks in the energy sector.
Decision 1364/2006/EC of the European Parliament and of the Council of September 6, 2006, laying down guidelines for trans-European Energy Networks (TEN-E Guidelines) and decision N° 1229/2003.
Directive 1229/2003/EC of the European Parliament and of the Council of June 26, 2003 laying down a series of guidelines for trans-European energy networks and repealing Decision No 1254/96. *Official Journal of the European Union* L, 176(15), 7.
Directive 2009/73/EC of the European Parliament and of the Council of July 13, 2009 concerning common rules for the internal market in natural gas and repealing Directive 203/55/EC. *Official Journal of the European Union*, 14.
European Commission, 2010. *Communication: Energy Infrastructure Priorities for 2020 and Beyond – A Blueprint for an Integrated European Energy Network*. COM(2010) 677 final. Brussels: European Commission.
European Commission, 2011. *Proposal for a Regulation on Guidelines for Trans-European Energy Infrastructure and Repealing Decision Number 1364/2006/EC*. COM(2011) 658 final. Brussels: European Commission. See http://eur-lex.europa.eu/LexUriServ/LexUriServ.do?uri=COM:2011:0658:FIN:EN:PDF.
Makholm, J. D., 2012. *The Political Economy of Pipelines: A Century of Comparative Institutional Development*. Chicago, IL: University of Chicago Press.

North, D. C., 1991. Institutions. *Journal of Economic Perspectives* 5(1): 97–112.
Talus, K., 2013. *EU Energy Law and Policy: A Critical Account.* Oxford: Oxford University Press.
Trombetta, J., 2012. *European Energy Security Discourses and the Development of a Common Energy Policy.* Working paper no 2. Groningen: Energy Delta Gas Research.
Williamson, O. E., 2000. The new institutional economics: taking stock, looking ahead. *Journal of Economic Literature* 38: 595–613.

4 Addressing energy security from Brussels

The merits of Regulation 994/2010

Introduction

The previous chapter laid out how member states in two elements of the gas system (i.e. infrastructure and regulation) continue to have a strong mandate, even though significant developments (ACER, energy infrastructure legislation) have taken place in recent years. In addition, it is important to keep in mind that in the larger part of the EU investments in infrastructure have been made. Similarly, the Lisbon Treaty has been hailed for finally making energy security a European issue, though arguably article 194 continues to be explicit in that member states have the final say when it comes to choices regarding measures that address energy security, or the fuel mix.

The EC attempted to address energy security long before the Lisbon Treaty however. Directive 2004/67/EC was seen as the "first legal frame work at Community level to safeguard security of gas supply and to contribute to the proper functioning of the internal market in case of disruptions."[1] As is discussed below, Directive 2004/67/EC was succeeded by Regulation 994/2010, concerning measures to safeguard security of gas supply. These policies are discussed in detail in this chapter. Subsequently it is assessed to what extent these policies have contributed to energy security in the member states. Of course, this is somewhat complicated to establish, because since 2010 the EU has not faced a supply disruption of noteworthy proportion. In June 2014, however, when Gazprom decided to cut gas supplies to Ukraine because of an open debt of several billion dollars (the exact amount is disputed and as of writing under review of the Arbitration Institute of the Stockholm Chamber of Commerce), there was broad concern among European policy makers that in the winter of 2014 supplies to Europe might be halted as well. Subsequently EC policy makers carried out large scale "stress tests," in which several scenarios of (partial) supply disruptions were modelled to see whether – theoretically – EU member states would be harnessed against these disruptions. As such, one could argue that these tests formed the first Litmus test of the measures taken at the member state level, and coordinated at the European level, to improve energy security. This chapter then concludes with a brief discussion of the merits of these policies.

When publishing Regulation 994/2010 the EC argued that security of gas supply mainly had to be a European concern, because it "cannot be sufficiently achieved by the member states alone and can therefore, by reason of the scale or effect of the action, be better achieved at the Union level."[2] The implementation of the regulation had to be monitored by a competent authority that member states had to appoint before December 3, 2011.

Towards preventive action plans and emergency plans

Directive 2004/67/EC was written well before Europe suffered from a supply disruption, like it did later in 2006 and 2009. Hence, in the context of stable supplies from external suppliers EC policy makers noted that natural gas as a fuel source was becoming more important, and that some minimal coordination between member states was required regarding an approach to security of supply, in order to avoid distortions of internal market functioning.[3] Member states were summoned to clearly define the roles and responsibilities of the different market players in order to achieve security of supply in specific situations (e.g. extremely cold temperatures, supply disruptions in which the EU risks losing more than 20 percent of its external supplies, or periods of exceptionally high gas demand that statistically occur every 20 years).[4] Arguably the most important elements of the Directive, however, concerned reporting requirements, and the installation of the Gas Coordination Group. Article 5 of the Directive built on the Second Gas Directive (2003/55/EC) and summoned member states to include in their reports to the EC the levels of storage capacity, information on long-term gas contracts, including their remaining duration but excluding commercially sensitive information, and the regulatory framework to provide incentives for new investments in domestic gas exploration, storage, LNG facilities, and gas transmission and distribution infrastructure. The Gas Coordination Group consisted of industry representatives and policy makers that discussed and coordinated national measures to address security of supply.[5] In retrospect it seems fair to say that Directive 2004/67/EC was rather informal, and obviously by excluding commercially sensitive information the EC would only get its head around a part of the supply picture. Noël (2010) concluded that Directive 2004/67/EC was "notoriously benign."

In 2010, shortly after the publication of the Third Package and the second major supply disruption in 2009, the EC launched ideas for Regulation 994/2010. The proposed document raised substantial concern, both within the industry community (where in particular countries that produced significant volumes of natural gas, such as in the Netherlands, but also countries that had a diverse and secure gas supply portfolio, had strong reservations) and among scholars. Noël and Findlater (2009) observed a legitimate concern on the EC side that this Regulation would truly "bite," contrary to its predecessor. The main concern was that imposing a standard on supply security would likely leave some member states under-insured, while massively over-insuring others. Upon viewing the proposed regulation, Noël (2010) concluded that the regulation contained a number of flaws, and that because of these there would be serious challenges to get the regulation, though hailed for its intention, through the EU decision-making process. In essence the approach chosen by the EC focused on setting minimum standards of security of supply. Arguably, levels of security throughout the EU are arbitrary, and the costs of attaining a certain level of security are country specific (*ibid.*: 4). Moreover gas security is not a pan-European public good, and member states that are already "secure" (arguably a substantial majority of them) had no incentive to support ambitious standards.

At the time there was significant opposition from industry, and the last major natural gas producer within the EU, the Netherlands, against the regulation. However, in the winter of 2009 the second major supply disruption crippled daily life in several European member states, and some in the wider European Energy Community, such as Serbia. In Bulgaria, for example, where natural gas has a very modest market share in terms of primary energy consumption, with not even 3.5 bcm of annual consumption at the time, the gas system could not effectively deal with the loss of almost 50 percent of its contracted natural gas. Though electricity generation was not meaningfully impacted, natural gas is used predominantly for heating and industrial activity, and it took authorities in Bulgarian capital city Sofia over one week to shift to heavy fuel oil for heating (Silve and Noël 2010). In light of the evident inability of several European member states to adequately respond to a major supply disruption, Regulation 994/2010 was eventually adopted in October 2010.

Regulation 994/2010 is designed to safeguard security of gas supply by allowing for exceptional measures in case the market can no longer deliver the required gas supplies. It also defines security of gas supply as a shared responsibility of natural gas undertakings, member states, and the EC.[6] The EC proposed to closely involve the Gas Coordination Group and to establish Preventive Action Plans and Emergency Plans.[7] These plans installed infrastructure standards, for instance on necessary capacity to satisfy total gas demand, made public before December 3, 2012. Moreover, member states are obligated to adhere to a certain infrastructure standard, and to make an assessment whether this standard is fulfilled at the national or regional level. Also, all cross-border interconnectors are expected to have bidirectional capacity (meaning natural gas can flow both ways) no later than December 2013.[8] Article 7 of the Regulation does provide a number of scenarios in which member states can be exempted. Based on all this information and supply standards under exceptional market conditions, emergency plans had to be designed. These identified what non-market based measures could be used in the case that market undertakings could no longer guarantee security of supply. The regulation also stipulated three main crisis levels:

- early warning (indication but no real disruption);
- alert (disruption but the market can take care of it); and
- emergency (despite all market intervention supplies can still not meet demand and additional measures are required).

Surely the announced measures provided the EC with a lot of market data. To give an example, the competent authority had to collect on a daily basis the gas supply and demand forecasts and the flow of gas at all cross-border entry and exit points connecting a production facility, a storage facility or an LNG terminal to the network.[9] Subsequent to these requirements, the EC had also decided that in order to assess the status of security of supply at the EU level it needed all relevant information of existing and future inter-governmental agreements with third countries, to be delivered to the EC by the competent authority in aggregated form.[10] Thus, the EC seemed to cut in on the action when security of gas supply was concerned. Examining its considerations that introduce the regulation in combination with considerations from earlier directives on the internal energy market, one can observe that the EC had slowly lost its faith that European market forces alone would guarantee security of supply in the EU, in case of emergencies.

Regulation 994/2010 was barely the end of the European debate on security of gas supply however. De Jong *et al.* (2012) suggested that although it was another step in the right direction, in essence Europe still had not achieved a European approach to ensuring gas supply security. In 2012 the EC Joint Research Center published a review of member states preventive action and emergency plans (Zeniewski *et al.* 2012). They noted that emergency plans had already been drafted by most member states prior to the adoption of Regulation 994/2010. The novelty were preventive action plans, which typically relate to the construction or upgrading of gas infrastructure, activities that are also carried out under different mechanisms such as ENTSO-G's Ten Year Network Development Plan that was discussed in the previous chapter. What stands out in the analysis, however, is how different member states interpreted Regulation 994/2010, which in essence should have been transferred directly into national legislation directly. Zeniewski *et al.* noted, however, that both procedures and criteria under which crisis levels were declared differed per member state. To give an example, France only has one crisis level, whereas the UK and Ireland have five (the regulation mandates three). Also, the assessment whether there is a crisis level is sometimes made by a government actor (France), a TSO (Belgium), a designated crisis manager (Ireland), or a regulator (Germany) (*ibid*.: 22). Then, member states have different approaches when it comes to utilizing non-market based interventions. Austria, for example, can only issue these measures when all other measures have failed, whereas in Poland the Minister of Economy can intervene when the second level (of four) of a crisis has been entered. The analysis also showed that theory and practice of suggested measures did not necessarily align. To give an example, Latvia could in theory supply neighboring Lithuania with stored natural gas. The technical capacity for natural gas withdrawal from storage is 5 mcm per day, but due to limitations on the Latvian transmission network it can only ship 20 percent of that (*ibid.*: 35). The overarching and most important conclusion of the report, however, was that in most cases under study emergency plans treat supply disruptions as exogenous and focus on domestic response options. Few national policy makers have thought of increasing imports of natural gas from neighbors as a viable strategy to deal with a supply disruption, and there was little emphasis in the preventive action plans on building resilient networks that take a regional, rather than national focus. There were even examples where emergency policies prohibit transnational action: in Slovakia during emergency situations operators of storage facilities are obliged to stop supplying foreign customers. Belgium has similar policies in place (*ibid.*: 45, 46). And in defense of countries taking a national approach to address energy security, there are examples where a national approach under current circumstances provides more safeguards, even though that goes at the cost of neighbors. If Greece were to cooperate with neighboring countries and help send some of the LNG it can attract, after a 6 month supply disruption it would be worse off than would be the case when it would just keep the LNG in the country.[11] Hence, without sufficient infrastructure and reverse flow options, in individual cases a soloist approach, though undesirable from a European perspective, is understandable.

Looking at all these data, the question is valid whether Regulation 994/2010 had improved or further contributed to security of gas supply in Europe. Noël (in Kalicki and Goldwyn 2013: 183) concluded that the regulation would not be more effective than the directive it replaced "despite its extreme complexity and apparent sophistication." However, Talus (2013: 103) is more optimistic and concluded that the regulation has, contrary to the directive it replaced, created a "European approach" to gas security, and shifted the central role from the member state level to the European level. While in theory that case can be made, the data provided by the JRC indicated that remarkably little coordination took place at the member state level, and that perchance more surprisingly the Regulation seemed to be interpreted differently throughout the EU as well. What is more, since 2009 no disruption of natural gas supply has taken place in the EU.[12] Arguably in 2014 the perceived risk of a supply disruption, in light of the crisis in Ukraine, provided another test to see whether Regulation 994/2010 had contributed to improving energy security.

A first litmus test? "Stress tests" during the Ukraine crisis

Quarrels between Russia and Ukraine continued over the course of 2014, and one of the issues of dispute was an open debt of several billion dollars for natural gas that had been delivered to Ukraine in 2013. As negotiations proved to be fruitless, in early June 2014 Russia decided to seize all supplies to Ukraine until an agreement had been reached. Thus, on June 27 the

European Council endorsed a proposal by the EC to carry out what it called a stress-test exercise in order to test the resilience of European gas markets to disruption of gas supplies in the upcoming winter.

In October 2014, ENTSO-G and the IEA presented their findings to the EC.[13] In essence, it showed that resilience of the EU as a whole had increased substantially since 2009, when the last supply disruption took place. It also showed, however, that in the case of a six-month supply disruption, EU member states would still be a few percentage points of natural gas demand short. This already assumed replacing large amounts of Russian natural gas with LNG available on the spot-market, intense storage withdrawals and voluntary demand reduction. This does not sound all too dramatic from an overall EU perspective, but it is important to note that this is a modelling exercise that presumes perfect collaboration between all market actors and optimal utilization of existing infrastructure and available resources. In reality, none of that happens. It was therefore helpful that ENTSO-G had studied two scenarios, one of which was uncooperative, meaning there was no equal relative burden sharing, as was the case in the cooperative scenario. The outcomes were very different, and probably much more realistic. Towards the end of a six-month disruption period, supply shortfalls anywhere between 40 percent and 100 percent could materialize in Bulgaria and Romania (and Serbia, Macedonia, and Bosnia Herzegovina).[14] In the case that Russia would cut all supplies to Europe similar shortfalls would apply for Lithuania, Estonia, and Finland. In addition, Hungary and Poland would be affected by shortfalls of 30 percent and 20 percent, respectively.

The EC study contained two important overall conclusions. First, several infrastructure projects that were launched after the supply disruption of 2009, with the explicit purpose of increasing gas supply security, have not been commissioned.[15] The reasons mentioned in the report are a lack of political support, unsatisfactory project management, and a lack of cross-border cooperation. Second, many of the national gas supply strategies are unilateral, not coordinated sufficiently, or just insufficiently cooperative. Nevertheless, the EC concluded that the application of Regulation 994/2010 has resulted in "clear improvements" in the supply situation in comparison to 2009, though it believed that there were "margins for strengthening" the regulatory framework.

The EC also published a more elaborate working document which contains the findings of an in-depth examination of the implementation of Regulation 994/2010 and its contribution to solidarity and preparedness for gas supply disruptions in the EU.[16] The assessment looked in particular at five key elements from the regulation, namely the supply standard and protected customers, the infrastructure standards (N-1 rule), the Risk Assessment, Preventive Action Plan, and Emergency Plan, the notification of intergovernmental agreements and details of commercial agreements, and finally responsibilities in case of emergencies. Table 4.1 lists the key findings from this assessment.

Table 4.1 Main conclusions from the report on the implementation of Regulation 994/2010

Key element from Regulation 994/2010	Implementation and overall assessment
Supply standard and protected customers	Large discrepancies continue to exist and fundamentally different concepts in MS regarding the implementation of the supply standard. In several cases methodology for controlling and enforcing the implementation of the supply standard is missing. Very often basic information to verify the fulfillment of the standard is missing, or there is a lack of detail or precision thereof. More generally, it appears that "Regulation 994/2010 has failed to bring about a clear system in which the supply standard is monitored and enforced in a systematic manner." As a result citizens throughout the EU remain unequally protected.
Infrastructure standard (N-1 rule and obligation to install bidirectional capacity)	MS complying with the N-1 rule has increased to 20, and 3 MS are exempted. The situation in Bulgaria remains highly problematic, and in the short term Greece, Lithuania and Latvia are also vulnerable. Compliance with the N-1 rule has to be combined with other indicators such as flexibility, otherwise it can give a false impression of security. Compared with 2009, four more borders have become bi-directional (Germany–Denmark, Italy–Austria, Greece–Bulgaria, Romania–Hungary). Regulation 994/2010 has been instrumental in putting in place or speeding up physical reverse flows on some interconnectors (i.e. Poland–Germany, Hungary–Romania, Greece–Bulgaria) even though the majority of these improvements have been incentivized by market demand. Though several MS such as the Netherlands, Germany, and Czech Republic fulfill the N-1 rule, additional bi-directional flows in these countries would better facilitate natural gas from the west of Europe to flow eastward.
Risk assessment, preventive action plan, and emergency plan	The main weakness is that the plans are almost exclusively nationally focused. The cross-border impact of national measures is not taken into account.
Notification of intergovernmental agreements and details of commercial agreements	The obligation regarding intergovernmental agreements became obsolete with the adoption of Decision 994/2012/EC.[a] Since commercial information had to be handed in on an aggregate level, the value of this information has overall decreased.
Responsibilities and coordination in case of emergency	The high level of transparency regarding cross-border capacities, underground storage levels, and gas flows enables rapid action in case of anomalies.

Note: [a] This decision established an information exchange mechanism with regard to intergovernmental agreements between member states and third countries.

Source: SWD(2014) 325 final – Commission Staff Working Document

Discussion

Even though an actual supply disruption has not taken place since the adoption of Regulation 994/2010, it is safe to say that its contributions to European energy security are modest. To suggest the formulation of one standard of energy security in all the member states was probably not realistic, if only because (perceptions of) security differs per member state. So do the various levels of import dependence. What is more, over four years after the adoption of Regulation 994/2010 it is remarkable to see that upon evaluation, in several cases basic information is still lacking, ironically often in member states that are the most vulnerable in terms of market manipulation and that sometimes are single-source dependent. This raises questions, for instance to what extent energy dependence is a real security concern among national policy makers. Despite all the rhetoric, are we dealing with a real problem here? It seems questionable, at least to an extent, because if these risks were as grotesque as often portrayed, then it does not make sense that legislation aimed to improve the situation is not implemented swiftly, and sometimes not implemented at all. Or is it ineffective policy, and also perceived as such? There is no evidence to support that notion at this point.

Positive features of Regulation 994/2010 arguably are an increased level of information and transparency. This helps policy makers, both at the national and at the supranational level, better understand where bottlenecks and vulnerabilities are in the EU gas system. For EC policy makers, this helps coordinate action to address these vulnerabilities, even though the instruments available to do so are relatively modest, as has been discussed in Chapter 3. What follows from the analysis and "stress-tests" as carried out by the EC is that energy security in essence continues to be an issue that is framed with an almost entirely national perspective. In that sense the concept of energy security has not changed, and the question is valid whether it will.

In sum, Regulation 994/2010 so far has contributed to increased transparency and the overall level of information. On the other hand, member states have failed to implement what are basic notions of the legislation. Therefore, the contribution of the regulation in terms of European energy security is modest at best. Surely, Talus (2013: 108) is correct when he states that the criticism that the EU proceeds too slowly is based on incorrect assumptions as to how a continent-wide entity should proceed. Similarly, Aalto and Temel (2014) observed that at best a limited degree of further integration is likely. Yet it is worth emphasizing that in combination with – sometimes fierce – rhetoric about dependence on Russian supplies, the lack of progress when it comes to implementing fairly basic legislation to address these issues, is striking (see also Noël in Kalicki and Goldwyn 2013: 172). In defense of those that critique the EU for not moving fast enough, the tangible actions in several member states to address energy security are so slim that there arguably must be more to it than just a misunderstanding of EU integration and collaboration dynamics.

Notes

1   Regulation 994/2010.
2   *Ibid.*, consideration 49.
3   Directive 2004/67/EC, consideration 3.
4   *Ibid.*, articles 3 and 4.
5   *Ibid.*, art. 7.
6   Regulation 994/2010, articles 1 and 3.
7   To underline the importance of energy security for the European institutions, see for instance the EP report on the implementation of the European Security Strategy and the Common Security and Defense Policy (2009/2198 (INI)), A7–0026/2010.
8   Regulation 994/2010, art. 6.
9   *Ibid.*, art. 13 sub 2.
10 *Ibid.*, art. 13 sub 6 a + b.
11 SWD(2014) 326 final – Report on the findings of the South-East European focus group, p. 4.
12 Until November 2014 that is.
13 COM(2014) 654 final Communication on the short term resilience of the European gas system.
14 *Ibid.*, p. 6.
15 *Ibid.*, p. 15.
16 SWD(2014) 325 final – Commission Staff Working Document.

References

Aalto, P., Temel, D. K., 2014. European energy security: natural gas and the integration process. *Journal of Common Market Studies* 52(4): 758–774.
De Jong, J., Glachant, J.-M., Hafner, M., Ahner, N., Tagliapietra, S., 2012. A new EU gas security of supply architecture? *Policy Brief* 2012/03, June See http://cadmus.eui.eu/bitstream/handle/1814/22500/PB_2012_03_digital.pdf?sequence=1.
Directive 2003/55/EC of the European Parliament and of the Council of June 26, 2003 concerning common rules for the internal market in natural gas and repealing Directive 98/30.
Directive 2004/67/EC of April 26, 2004 concerning measures to safeguard security of natural gas supply.
Kalicki, J. H., Goldwyn, D. L., (eds) 2013. *Energy and Security: Strategies for a World in Transition*, second edition. Washington, DC: Woodrow Wilson Centre Press.
Noël, P., 2010. *Ensuring Success for the EU Regulation on Gas Supply Security*. Cambridge: Electricity Policy Research Group. See www.eprg.group.cam.ac.uk/ensuring-success-for-the-eu-regulation-on-gas-supply-security.
Noël, P., Findlater, S., 2009. *A Comment on the Draft Regulation on Gas Supply Security*. Cambridge: Electricity Policy Research Group. See www.eprg.group.cam.ac.uk/a-comment-on-the-draft-eu-regulation-on-gas-supply-security.
Regulation No 994/2010 of the European parliament and of the Council of October 20, 2010 concerning measures to safeguard security of gas supply and repealing Council Directive 2004/67. *Official Journal of the European Union*, 53(L295).
Silve, F., Noël, P., 2010. *Cost Curves for Gas Supply Security: The Case of Bulgaria*.

Cambridge: Electricity Policy Research Group. See www.old.cambridgeeprg.com/?papersonly=1&s=Cost+Curves+for+Gas+Supply+Security%3A+The+Case+of+Bulgaria

Talus, K., 2013. *EU Energy Law and Policy: A Critical Account.* Oxford: Oxford University Press.

Zeniewski, P., Bolado, R., Gracceva, F., Zastera, P., Eriksson, A., 2012. *Preventive Action Plan and Emergency Plan Good Practices: A Review of EU Member State Natural Gas Preventive Action and Emergency Plans.* Report EUR 25210 EN. Petten: Joint Research Centre, Institute for Energy and Transport.

Part III
Case studies

5 A structural lack of investments in natural gas infrastructure

Introduction

The challenge to interconnect and adapt European energy infrastructure is significant and urgent. The EC estimated in June 2011 that the total investments needed in energy infrastructure up to 2020 were roughly €200 billion. These estimates have been confirmed repeatedly, most recently in 2014, although details have not been published. Assuming that natural gas continues to play a crucial role in the EU energy mix, about €70 billion of that total amount would be needed for investments in gas transmission infrastructure, storage facilities, liquefied natural gas (LNG) terminals and reverse flow capacity. The EC also estimated that these necessary investments would not take place under business-as-usual conditions, because of problems related to permit granting, regulation and financing (European Commission 2011). With regard to regulation, the EC considered the existing framework to be "not geared towards delivering European energy infrastructure priorities."[1] Scholars have analyzed the effects of regulatory uncertainty in western Europe (that is generally believed to have a reasonably well developed regional gas market) and found that under current tariff policy and with uncertainties regarding demand and supply, it is unlikely that market forces will attract sufficient investments in gas transport capacity (Pelletier and Wortmann 2009). In view of the desired levels of security of supply and servicing new (sustainable) energy resources, other parts of the continent, most notably central and eastern Europe, have significantly less developed gas systems and are currently under construction (see European Commission 2012; Johnson and Boersma 2013). On the other side of the Atlantic Ocean, in the US gas system, it has been argued that there are no reasons for concerns about investment in gas infrastructure (Von Hirschhausen 2008).

As is described in more detail in Chapter 7, some scholars have looked at the US gas system in order to draw lessons from it (e.g. Ascari 2011). It is worth noting that there are some fundamental differences between the two systems, one of them being the lack of institutional history in the EU in terms of energy market liberalization – which is currently in transition towards one internal gas system – in contrast to the US, where this process first started in 1938 with the passage of the Natural Gas Act, giving the federal government direct involvement in the regulation of interstate natural gas (for a detailed account of this institutional history, see Makholm 2012).[2] Nevertheless, the (overall) lack of concerns about investments in gas infrastructure in the US asks for further examination, given the significant and urgent challenges that the EU faces. The necessity to address this issue was once again demonstrated in light of the 2014 Ukraine crisis. Analyses, or so-called "stress tests," published by the European Commission clearly indicated the continued lack of market integration in parts of the EU gas system, which makes several member states vulnerable to supply shocks and abuse of market power by the dominant supplier (European Commission 2014a). Therefore, despite the myriad of differences between the two gas systems, this chapter aims to investigate them by making an institutional analysis of the EU, using the case of the US as a benchmark. Hence, the main question in this case study is: What lessons can be learned from the investment climate regarding gas transmission infrastructure in the EU and the US, drawing from an institutional analysis? Although there is not a single regulatory approach within the EU, Jamasb *et al.* (2008) have argued that most member state regulatory authorities follow the examples set by Great Britain and the Netherlands, and therefore this study focuses on these two cases. It is important to note that these cases give us some but limited information about EU regulatory approaches in general. Ruester *et al.* (2012b) observe that across Europe various forms of general price-control mechanisms co-exist (e.g. rate-of-return, price-cap, and revenue-cap regulation). Second, there is wide heterogeneity with regard to the calculation of allowed revenues, and finally, there are considerable differences in regulatory period. The chapter deals mainly with existing decision-making structures in the EU and the US. In terms of regulatory regimes and regulatory instruments, it offers the basic ideas to outline the fundamental differences between the two continents from a regulatory perspective. A detailed regulatory economic analysis of these systems is beyond the scope of this book.

This chapter starts with an outline why the European gas system's status quo is suboptimal. Most of the examples used in this section come from the Netherlands. The chapter continues by describing basic characteristics of the US gas system. Both sections provide a brief state-of-the-art overview of relevant academic literature. Subsequently the chapter describes the relation between legislature and regulatory authorities and their mandates in the EU, in the individual member states and in the US, and discusses these relations. Then, criteria are analyzed that regulatory authorities apply when determining gas transport tariffs. Hereto the broad energy policy goals as laid down in the Lisbon Treaty are used (i.e. security of supply, efficiency, and sustainability). These criteria are applied to the US case as well. As the analysis shows, not every regulatory authority uses all of these policy goals.

In an attempt to quantify these policy goals, efficiency is examined by look-ing at what rates of return regulatory authorities allow gas infrastructure operators to make. Quantification of the other policy goals is more chal-lenging. Security of supply can be measured by collecting data on interruptions in gas flows, but those are not always available and interrup-tions can have multiple causes. Sustainability through regulation could be measured in terms of net CO_2 reductions, but studies like that have not been carried out thus far. Next, this chapter focuses on the role of both private and public investments in gas infrastructure and what lessons can be derived from differing existing practices and the ongoing academic debate regarding this topic. Finally, decision-making structures in both cases are analyzed, using a multilevel governance (MLG) framework.

Both primary and secondary data, covering the period up to December 2014, are used in this chapter. The former is primarily qualitative and derived from interviews with representatives of regulatory authorities and infrastructure companies. The latter consists mainly of academic contribu-tions, policy papers and legal documents.[3]

Why the European Union status quo is suboptimal

First, drawing from the discussion on EU energy security in Chapter 2, the academic debate on this particular topic has mostly been focused on limited resources or unreliable external suppliers. The organization and functioning of the EU internal energy system, in particular regarding regulation and energy infrastructure, is on the whole less extensively explored. Research on energy regulation appears to face that same pitfall by focusing mainly on efficiency. As an illustration, Viscusi *et al.* (2005: 9) wrote that "Ideally, the purpose of antitrust and regulation policies are to foster improvements judged in efficiency terms." However, energy regulation is part of a broader policy area, also including issues such as safety, security of supply and sustainability. The legislature often serves multiple and sometimes changing policy goals, which can go beyond and/or against establishing efficient investment decisions and dealing with antitrust issues. Kwoka and Madjarov (2007: 26) stated that "economic theory explains the way to maximize efficiency, whereas other societal objectives could not be achieved by competitive markets." Furthermore, transmission system operators may be required by law to carry out tasks that are not explicitly part of the regu-latory framework. To give an example, European transmission system operators are required by law to make provisions safeguarding security of supply, while at the same time European regulatory authorities commonly do not, or at least hardly, use security of supply as a criterion when assess-ing proposed transport tariffs.[4]

Second, the current fragmented organization of the EU energy system is not always efficient. An example hereof is the different tariffs and access conditions that infrastructure companies are allowed to introduce for the transport of natural gas in the Netherlands and Germany. In the Netherlands, since 2004 there has been a single gas transmission system operator, named Gas Transport Services (GTS).[5] This is a wholly owned subsidiary of Dutch Gasunie, a private limited liability company, in turn owned (100 percent) by the Dutch state but required by law to act independently.[6] In July 2008, GTS purchased two transmission networks in northern Germany, BEB and EMGTG, respectively from Shell and ExxonMobil. This extended the network of GTS to Berlin, supposedly providing it with a strategic position regarding the Nord Stream pipeline that runs from Russia through the Baltic Sea to northern Germany. Shortly after the purchase, in 2009 the German regulatory authority BundesNetz Agentur, tasked with administering and approving the tariffs calculated for the transmission of natural gas, decided to lower the maximum turnover to be achieved by gas infrastructure companies in Germany. As a result, GTS had to devalue €1.52 billion of its initial purchase of €2.15 billion.[7] This resulted in a political debate in the Netherlands regarding the spending of public financial means in risky purchases abroad, and moreover, as anecdotal evidence suggests, the earlier adverse investment has negatively influenced the possible purchase of the adjacent gas network of German multinational company Thyssengas Netz (RWE), which had been put on sale at the end of 2008 under pressure of the EC.[8]

Third, despite steps that have been taken to complete the internal market for natural gas, barriers to competition remain. Gasmi and Oviedo (2010) noted that these are mostly related to market structure, national attitudes towards liberalization, access to gas supplies and access to key infrastructure facilities. Regarding the latter, there appears to be growing consensus among European policy makers that the existing regulatory framework is not fit to address the major energy infrastructure needs in the decade ahead. This idea is based on the notion that some infrastructural investment projects are not taking place because they provide higher regional than national benefits (e.g. costs have to be made in one member state, while benefits are enjoyed by several neighboring member states), while others do not because they use innovative technologies with higher risks and uncertainties. Ruester *et al.* (2012a) conclude that current EU involvement in the regulation of TSO revenues and transmission grid tariffs is rather limited and that the existing heterogeneity among regulatory practices may be an obstacle for adequate investments in the grid. In addition, there are projects with externalities (i.e. impacts), which are not taken into account by market demand, because they are disregarded in the investment decision (European Commission 2011).[9] Regulatory authorities currently focus mainly on efficiency when setting tariffs for the transport of natural gas. This is remarkable given the broader policy agenda that looms in the background.[10] It is worth investigating to what extent regulatory authorities consider that wider agenda when designing tariff structures, or taking other regulatory measures, or whether these are otherwise part of the mandate of regulatory authorities under study.

Fourth, relations between the national legislature and regulatory authorities are not always clear. At the request of the EP, the Dutch regulatory authority (and its colleagues in other member states) was granted the judicial status of an autonomous administrative authority in 2005.[11] The goal of creating this status was to confirm the importance of having an independent regulatory authority in the energy sector and thus depoliticize its work. In 2010, this status caused remarkable friction when the Dutch Trade and Industry Appeals Tribunal (CBb) annulled the tariff regulation that the Dutch transmission system operator GTS had designed in accordance with Dutch government policy. The then Minister of Economic Affairs laid down several conditions that GTS had to apply when designing tariff structures, such as the value of the national infrastructure system, the terms of depreciation and the remuneration of the cost of capital. However, the Council for the Judiciary stated this was unlawful since the Minister was meddling with the competences of the independent regulatory authority.[12] For that reason, in 2010 the Dutch regulatory authority proposed a new method of regulation for the period starting in 2006.[13] While these tariffs were published and a "settlement deal" of €400 million was proposed by the regulatory authority, both GTS and the joint major industrial consumers indicated that they were not satisfied. This resulted in a new legal battle that ended in November 2012, in favor of GTS. All in all the process has created considerable uncertainties and setbacks for the infrastructure company and most likely negatively affected the development of new similar business ventures in the Netherlands.[14]

Fifth, in the Netherlands there has been an ongoing political debate about whether gas networks should be (partly) privatized. GTS's limited financial clout to make necessary investments may have fuelled this debate. The request to unbundle vertically integrated energy companies, driven by the EC from the late 1980s, has in the Netherlands always been answered for by the argument that networks should in principle be in public hands, because the secure and stable access to energy is an important public concern. Hence, in a liberalized European energy market, the only reasonable thing to do was unbundle the integrated companies. Several years later the unbundled system operator GTS itself appears to be one of the advocates of the privatization of a minority share of the Dutch transmission system, allegedly because it needs additional funds to make the required investments. In addition, in June 2011 an EC staff working paper stated that "Investors, such as public banks or investment funds, confirmed that transmission system operators have largely exploited their ability to raise debt capital and that future investments will require large equity injections by private investors or the State" (in case of publicly owned transmission system operators).[15] Examples like these demonstrate the need to examine privately funded gas networks, by for instance looking at the US or Great Britain, where private investments in gas networks are common practice.

Natural gas infrastructure in the United States

After three decades of regulatory reforms the US natural gas system has moved from a highly regulated to a highly competitive industry.[16] Siliverstovs *et al.* (2005: 613) argued that the European natural gas market is also highly integrated, but that the real issue is that natural gas markets across the Atlantic Ocean are not integrated, leaving gas prices in the US to be determined in a more market driven manner, while European gas prices follow the oil-linked model. Yet other and more recent studies dispute that the European market is integrated and hint at only partial integration (i.e. in northwestern Europe; Renou-Maissant 2012; Asche *et al.* 2013). It makes sense to point at the US/EU difference of literally thousands of producers of natural gas in the US (that one would not find in Europe) seeking a market that brought on the development of a spot market in the 1980s (Herbert and Kreil 1996). Others have also attributed part of the contended "successes" of US liberalization of the gas market to infrastructure institutions and regulation (Medlock 2012). De Vany and Walls (1994) argued that before institutionalizing open access to gas pipelines, the necessary regulation basically fragmented the natural gas industry and turned the pipeline grid into islands. Gas would only flow where long-term contracts between gas field and buyer told it to flow. Removing that barrier created dozens of gas spot markets that became highly integrated within two years of open access (*ibid.*: 757). Makholm (in Lévêque *et al.* 2010) allocated even more importance to the role of pipeline and regulation reforms in the US liberalization of gas markets' acclaimed success, stating that the "answer to gas security lies in pipelines." He concluded that more flexible and transparent transport systems, increased flexibility in supply contracts, moving away from oil-indexation and lower costs for network usage should be objectives of European policy makers (*ibid.*: 49).

Although scholars have pointed at the benefits of competition in the US natural gas system in terms of more efficient production, they also expressed concerns over increasingly volatile prices due to demand shocks (Mohammadi 2011). In addition, while there is substantial agreement that US natural gas markets are largely integrated, empirical evidence suggests that there are considerable differences in the degree of integration of the individual hubs into the national market, indicating at least temporary or short-term market power at four of the nineteen studied trading hubs (Murry and Zhu 2008). Heather (2012) predicted a similar scenario (with mature and less developed trading hubs) for Europe. Another distinguishing factor of the US natural gas system is competition in infrastructure. While the details of this feature fall outside the scope of this chapter, it is worth noting that one of its consequences is that in non-densely populated parts of the US people are not connected to natural gas grids, because building a distribution network here would not be economical. In 2011 merely 65 million US residential customers were connected to gas distribution grids, a number that has grown to 68 million in 2014 (and more than 5 million commercial enterprises).[17] In parts of the country without gas networks, alternative fuels are used (e.g. distillate oil, propane, or wood). As an illustration, in the US in 2011 around 8 million residential customers used propane as an alternative fuel source.[18]

Nonetheless, the above raises the question whether – with the consideration of the caveats that have been touched upon – lessons can be learned from the case of the US. Moreover, an envisaged shift of policies and regulation towards a more competitive model for the EU gas market is observed, away from long-term and oil-indexed contracts and therefore more in line with the US gas market (see Box 5.1 on the Gas Target Model).

Von Hirschhausen (2008) concluded that in the restructured US natural gas system there is little reason for concern about infrastructure investments. In a case study that examined not only gas transmission pipelines but also LNG infrastructure and gas storage facilities, no evidence was found of underinvestment. Rather, on top of the argument regarding US rate of return regulation that it leads to inefficient use of capital and labor, he concluded that this same regulatory framework also secured long-term investment (*ibid.*: 7). Currently in parts of the US substantial investments in gas infrastructure are required, but these are linked to the very significant increase in domestic gas production in parts of the country. This is discussed in more detail in Chapter 6. As an illustration, the Interstate Natural Gas Association of America (INGAA 2014) estimated that investments in new natural gas transmission capacity (including pipelines, storage facilities, processing facilities and LNG export facilities) until 2035 hover around an average of US$14 billion per year, totaling US$313 billion.[19] In a similar study done in 2011, INGAA estimated that this annual investment requirement would hover around US$8 billion per year, which gives an indication of the size of natural gas production growth in the US, and also the substantial uncertainties that come with long-term forecasts. Notwithstanding these substantial investments, there is no empirical evidence that suggests that this is a structural problem. Boersma and Ebinger (2014) described how it takes between 30 and 45 days to move a drilling rig and start the production of natural gas elsewhere. In contrast, years are required to construct a pipeline. Therefore, and absent evidence that suggests a structural problem, it is uncertain whether proposed legislation to accelerate the decision-making process by the federal regulator (FERC) could solve the aforementioned mismatch between gas production and infrastructure construction. Surely, the infrastructure industry has not been able to keep pace with the rapid expansion of natural gas production in the US, but beyond short-term bottlenecks, more work is required to establish whether there is a long-term problem. Anecdotal evidence suggests that this is not the case. Jamasb *et al.* (2008) found that, taking productivity and convergence as performance indicators, regulation has been rather successful in the US in a data envelopment analysis of US gas transmission companies.

Box 5.1 CEER vision for a European Gas Target Model

In December 2011, the Council of European Energy Regulators (CEER) published its concluding paper on the Gas Target Model. The paper contains a vision on the future EU gas market and proposes measures to complete the internal gas market by 2014.

In their approach, regulatory authorities think of a competitive EU gas market as a combination of entry–exit zones with virtual hubs. Competition should be based on the development of liquid hubs across Europe where gas trade takes place. Price signals and efficient usage of infrastructure should facilitate gas to flow wherever it is most valued. Hence sufficient and efficient investment in infrastructure should be facilitated.

The status quo is slightly different: historically Europe has met security of gas supply through long-term contracts and facilitating storage of natural gas to provide seasonal and short-term flexibility. Though in the northwest of Europe wholesale trade on hubs has made progress, see for instance British hub NBP or Dutch TTF, CEER argues there is much work to be done. Some of this has been initiated through ACER (e.g. capacity allocation mechanisms and proposals for harmonized tariffs structures). CEER finishes its paper with three recommendations:

- The 3rd Energy Package in general and entry–exit systems in particular must be implemented as soon as possible. Assessment by the national regulatory agencies should be complete at the end of 2012.
- Capacity allocation mechanisms and congestion management proposals must be adopted and implemented by 1 January 2014 at the latest.
- CEER develops proposals how to identify and integrate new capacity through coordinated market-based procedures.

Source: www.energy-regulators.eu/portal/page/portal/EER_HOME/EER_CONSULT/
CLOSED%20PUBLIC%20CONSULTATIONS/GAS/Gas_Target_Model/CD/
C11-GWG-82-03_GTM%20vision_Final.pdf (accessed September 12, 2012).

Subsequently they concluded that bench-marking based regulation could be possible if data were available and moreover that in the long run, market integration and competition are alternatives to the European model. At the same time, Von Hirschhausen *et al.* (2004: 207) alerted to the risk of changing regulatory regimes while long-term contracts are still in place and the possibility of uncertainty leading to underinvestment. Hence they pleaded for a regulatory framework that balances various objectives, rather than focusing exclusively on the issue of investment in infrastructure.

Many of the aforementioned contributions touched upon the role of institutions in dealing with challenges related to gas infrastructure investment. However, so far they did not dissect how responsibilities regarding gas infrastructure are divided among institutions on both sides of the Atlantic Ocean. Rather, these studies focused on the way gas systems have been organized, and how historical developments, political decisions and regulation have shaped the gas system until the moment of analysis. Despite existing differences between gas systems in the US and the EU and arguably different states of development, current trends towards more competition, as identified in the EU, suggest that an analysis of the organization of these responsibilities may well provide useful insights into the successes and failures of organizing the gas market, and in particular generating investments in gas infrastructure. This chapter now turns to the relations between legislatures and regulatory authorities in both the EU and the US.

Relations between legislatures and regulatory authorities

Traditionally independent regulatory authorities were a US phenomenon, which only appeared in the EU in the 1980s and 1990s (Thatcher 2005). Subsequently a broad range of literature emerged about the reasons for politicians to delegate matters to regulatory authorities, the benefits of doing this and the drawbacks. The often-quoted reasons to create *independent* regulatory authorities are to enhance the credibility of policies and to shift complex technical issues to experts that are outside the political arena (see Elgie 2006). Thatcher (2002a) also suggested that creating independent regulatory authorities provides an easy route to shift possible blame for unpopular policies away from politicians. Larsen *et al.* (2006) observed that in the liberalization process of European energy markets, independent regulatory authorities were needed to increase the credibility of the process. This facilitated the need to separate the state as the owner of public utilities since it was also becoming the potential seller of those utilities (*ibid.*: 2867). In a survey of sixteen member states' electricity markets, they found that there is no correlation between the set-up of independent regulatory authorities and their choice of regulatory approach (*ibid.*). It is worth noting that this national approach of gas regulation is in contrast with the case of the US, where federal authorities have been directly involved in designing regulatory instruments since 1938. Several studies have also mentioned arguments against creating independent regulatory authorities, such as regulatory capture, the lack of accountability of these institutions as well as the lack of democratic legitimacy (*ibid.*; Christensen and Laegreid 2007; Maggetti 2009). Thatcher (2002b) concluded that these regulatory authorities have at least broken up what were previously private processes of regulatory decision making and thus made this process more transparent.

Formal delegation of powers is one thing; another is what this separation between legislatures and regulatory authorities means in practice. Some have argued that regulatory authorities are difficult to control by legislatures because regulatory authorities have access to information that is not available to legislatures and because it is very costly for legislatures to draft new policies to redirect regulation (Viscusi *et al.* 2005: 391). Maggetti (2009) observed that increasingly political power is delegated from democratic institutions to non-representative bodies that lack democratic accountability. He concluded that independent regulatory authorities play a central role in political decision making and they also increasingly play a key-political role in law-making (*ibid.*: 466). This is in line with Thatcher (2005) who concluded that politicians have in fact allowed regulatory authorities to become a distinct set of actors or "third force." Szydlo (2012) concluded that national regulatory authorities' economic and social goals often conflict, while at the same time European legislatures have shielded these regulatory authorities. He argued that depriving the member states' parliaments of the possibility to exert legislative influence on the activities of regulatory activities is dubious because it touches upon essential constitutional principles (e.g. the domain of the law). So, according to him, regulation of issues sensitive to citizens is an exclusive parliamentary prerogative (*ibid.*: 795).

The following sections analyze relations between legislatures and national regulatory authorities in the EU and subsequently the US. They also touch upon regulatory mandates in terms of the policy goals efficiency, security of supply and sustainability.

European Union

Within the EU the relation between national legislatures and regulatory authority in the gas industry is based on Directive 2003/55/EC. Member states are summoned to designate one or more competent institutions with the function of regulatory authority that "shall be wholly independent of the interests of the gas industry."[20] The driving force behind this clear separation was the EP. Regulatory authorities in the EU are responsible for two activities. First, monitoring and ensuring non-discrimination, effective competition and the efficient functioning of the market. Second, fixing or approving, at least the methodologies used to establish the terms and conditions for connection and access to national networks, including transportation tariffs and terms and conditions regarding balancing services.[21] Furthermore, member states may delegate the task to monitor security of supply to the regulatory authority.[22] Thus, regulatory authorities are considered to be technical units, operating independently from politics.

The third policy goal from the Lisbon Treaty, sustainability, is not mentioned in Directive 2003/55/EC. It is, however, the main subject of Directive 2009/28/EC on the promotion of the use of energy from renewable sources. Both legislature and transmission system operator have clear roles regarding investments in grids and grid codes.[23] Yet the mandate of the regulatory authority in terms of approving transportation tariffs, as described in the previous paragraph, does not apply as strictly here, for "if significant measures are taken to curtail the renewable energy sources in order to guarantee the security of the national electricity system and security of energy supply, member states shall ensure that the responsible system operators report to the competent regulatory authority on those measures and indicate which corrective measures they intend to take in order to prevent inappropriate curtailments."[24] The member states themselves have a decisive say in the investments needed when it comes to safeguarding sustainability related investments in infrastructure, as for instance indicated in article 16, sub. 4: "Where appropriate, Member States may require transmission system operators and distribution system operators, to bear, in full or in part, the costs referred to in paragraph 3."

United States

The approach of both legislature and regulatory authorities in the US is more market-oriented than the approach in the EU. This is underlined by the mission that the Department of Energy (DoE) pursues, namely "to ensure America's security and prosperity by addressing its energy, environmental and nuclear challenges through transformative science and technology solutions."[25] Basically, the US DoE relies on the market to safeguard sufficient energy supplies. The department's main activities are to gather statistics and fund research on the topics that are mentioned in its mission statement.

The Federal Energy Regulatory Commission (FERC) is the US regulatory authority responsible for interstate gas infrastructure. Comparable to the European case FERC is an independent regulatory agency, and can only be reviewed by federal courts. When focusing on the market for natural gas, the FERC deals mainly with interstate transmission pipelines and spends an estimated 10 percent of its time on intrastate pipelines (lines taking natural gas from transmission up to distribution level).[26] Similar to the US DoE, the FERC has adopted a laissez-faire approach, resulting in the application of a competitive regulatory model at transmission level (see Vazquez *et al.* 2012). The method and criteria used to determine transport tariffs and rates of return are examined in more detail in the next paragraph. Overall, a system operator can, together with market entities (e.g. shippers), submit a proposal for a new transmission pipeline at the FERC, which thereafter consults the market and the public about the intended project, controls the proposed rates and whether third party access is safeguarded, and subsequently approves of the project (or not). If, at any stage after construction, tariffs are changed (either due to construction costs, desire by shippers or because of other reasons) the FERC has to approve of these changes. If, however, all parties involved are satisfied with the status quo, the initial tariffs can remain unchanged, unless the FERC starts its own investigation (rate case). The basis for this competitive structure is found in 1992 Order 636 that aimed to "further the creation of an efficient national wellhead market for gas without adversely affecting the quality and reliability of the service provided by pipelines to their customers." This regulation, among others, required pipeline operators to unbundle their sales and transportation services, to provide access to storage facilities on an open access contract basis, open access transportation services that are equal in quality for all gas supplies, to offer all shippers equal and timely access to information relevant to the availability of their open access transportation service and to implement a capacity releasing program so that firm shippers can release unwanted capacity to those desiring it.[27]

In sum, while in the EU national regulatory authorities by law have a strong mandate regarding market functioning and setting the groundwork for transportation tariffs, monitoring security of supply and facilitating sustainability are matters that *may* be delegated to the regulatory authority. As is shown in Table 5.1, this in fact happened in the case of Great Britain. In the US, the federal regulatory authority mainly monitors market functioning by a laissez-faire approach and safeguards security of supply by allowing substantially higher rates of return (that are discussed later in this chapter) than its European counterparts, while sustainability is not part of its mandate and considerations.

It is worth considering to what extent the EP's position that regulatory authorities are merely technical units remains valid. The discussion about the different mandates suggests that the approach of regulation is (at least partly) based on political trade-offs. The three policy goals efficiency, security of supply and sustainability inevitably collide at some stage. It seems, for example, that in the US security of supply prevails as a policy goal, and it can be expected that this has consequences in terms of efficiency (more investments lead to higher costs). Also, EU regulatory authorities that contribute to the establishment of future transportation tariffs inevitably have to deal with trade-offs (e.g. Szydlo 2012). When for instance considering necessary future investments in infrastructure to facilitate the expansion of the share of natural gas with different caloric values (or "green gas" for that matter) the regulatory authority has to consider its consequences in terms of efficiency (rising transportation tariffs). Whereas regulatory authorities hence have crucial influence when it comes to attracting investments in energy infrastructure, and almost inevitably have to make decisions in terms of trade-offs, they are not politically responsible. As the example from the Netherlands earlier in this chapter suggested, *ex post* control over regulatory decisions can only take place in court. It is worth noting that from this perspective the EU and the US do not substantially differ, however, a major difference is found with regard to the level of decision making. In the EU regulation is primarily a national affair, but in

Table 5.1 Focus of British and Dutch regulatory authorities, in terms of Lisbon Treaty broad policy goals. Light grey shading indicates explicit mandate, black indicates the opposite, while dark grey indicates shared responsibility

Country	Efficiency	Security of supply	Sustainability
Great Britain	Delegated regulatory task	Shared responsibility with Department for Energy and Climate Change (DECC), and active engagement with UK government to carry out Gas SCR[a]	Delegated regulatory task for OFGEM since 2008
Netherlands	Delegated regulatory task	Shared responsibility with Ministry of Economic Affairs (though no explicit mandate) for the NMa,[b] which plays a role in design of new balancing regime	No delegated task for the NMa

Notes: [a] The Gas Security of Supply Significant Code Review is undertaken by OFGEM with support of the British government in order to determine whether reforms to the current gas balancing arrangements and/or enhanced obligations are required in order to improve security of supply.
[b] The NMa is the Nederlandse Mededingings Autoriteit, the Dutch regulatory authority.

the US regulatory measures regarding interstate gas transmission infrastructure are designed at the federal level. This chapter now turns to an overview of revenues that transmission system operators are allowed to make. More discussion about relations between regulatory authorities and legislatures follows in the final paragraph of this chapter, which deals with decision-making structures in both the EU and the US.

Revenues of gas transmission system operators

European regulatory authorities apply incentive regulation for the (unbundled) pipeline companies. Yet the FERC promotes competition through unbundling, flexible short-term rate setting, strong property rights and controlling the abuse of market power (Jamasb *et al.* 2008: 3399). Rate of return regulation prescribes a reasonable rate of return on investment for companies investing in infrastructure. One of the critiques is that it contains few incentives to operate efficiently (Viscusi *et al.* 2005: 436). Incentive regulation in theory is designed to create incentives for the regulated firm to lower costs, innovate, adopt efficient pricing practices and improve quality.

However, proper implementation is crucial, for the time path of the price cap must be independent of the firm's actual realized costs, so that efforts by the firm to lower costs do not automatically translate into a lower price (*ibid.*). In addition, some have argued that incentive regulation results in lower quality of service. This manifests itself in increased duration of service interruptions, not in increased frequency of interruptions (Ter-Martirosyan and Kwoka 2010: 260). Ruester *et al.* (2012b) caution that besides rates of return, other factors are equally important to stimulate for investments in grid infrastructure, such as regulatory stability, access to financial means, and transaction costs (e.g. due to long permitting processes). The next sections focus on the forms of regulation that are used in Great Britain, the Netherlands, and the US.

European Union

The Office of Gas and Electricity Markets (OFGEM) applies incentive regulation with price caps in Great Britain. The regulatory formula in short leads to an allowed revenue – derived from an estimate of the operating expenditure, capital expenditure, financing costs and taxes for the relevant period, together with the regulatory asset value – for the transmission system operator. For the regulatory period from 2007 to 2012, the allowed rate of return was (to the transmission system operator's satisfaction) 4.4 percent post taxes.[28] The regulatory asset value is one of the incentives that OFGEM uses to stimulate investments while at the same time reducing operation costs. This is elaborated on in the next section, when analyzing the incentives to stimulate investment in gas transmission infrastructure.

Next to safeguarding efficient tariffs, the tasks of OFGEM were recently expanded through new legislation. Starting in 2004, its mandate has also been to contribute to the achievement of sustainable development. This was confirmed in the 2008 Energy Act, which added to the duties of the Gas Markets Authority "the need to contribute to the achievement of sustainable development."[29] So how does this materialize on a daily basis? Based on the interviews with OFGEM representatives no examples have been found in the natural gas sector, but in electricity transmission one can think of incentives that were added to regulation in order to diminish the release of sulfur hexafluoride (SF_6) to the atmosphere because of its negative impact on the climate, and because this particular greenhouse gas is not covered under the European emissions trading scheme.[30] Furthermore, the regulatory authority publishes annual policy documents containing a sustainable focus of its activities.[31]

In the Netherlands, incentive regulation with price caps is applied as well. The Dutch NMa carries out an efficiency check, and, based on this assessment, determines the total revenues the transmission system operator can generate in each three-year period of regulation. The operator then uses these total revenues to propose transport tariffs for usage of its transmission pipelines.[32] The reasonable rate of return in the Netherlands is equal to the so-called weighted average cost of capital (WACC). The NMa calculates a bandwidth of real WACC values before taxes and subsequently averages the high and low values to determine the WACC. By doing this, the regulatory authority expects that the network operator receives the return it needs to operate efficiently, while at the same time expecting that this return will be representative for the whole period of regulation. Note that the WACC is gradually introduced, by means of a yearly correction of the maximum WACC value or cap (X-factor). Based on yearly performance the NMa applies incentives to stimulate efficient operations (X-factor) and safeguard quality standards that are required by Dutch law (Q-factor). For the current regulatory period (2010–2013) the real WACC before taxes is 5.8 percent, whereas this return was 6.5 percent in the period from 2006 to 2009.[33] The regulatory authority also advises the government regarding proposed investments by the TSO and plays a role in the design of market mechanisms such as the balancing system.

Throughout the EU different additional instruments have been used to stimulate investments in gas infrastructure (for a detailed overview, see Ruester *et al.* 2012b). Italy, for example, allows extra rates of return of 2–3 percent for a specified period of time for certain investments. In other cases, such as the BBL interconnector between the Netherlands and Great Britain, an exemption of third-party access rules, has been used to encourage investment (*ibid.*: 12). As discussed in detail in Chapter 3, public co-funding of infrastructure projects is possible, though financial means are limited in light of the required investment estimates.

United States

In contrast to the EU member states, the FERC applies rate of return regulation and has thus made a fundamentally different choice than its European counterparts (see Vazquez *et al.* 2012). As with incentive regulation, transmission system operators know up front what rate of return is allowed by the regulatory authority, but provisional correction mechanisms differ substantially. Whereas EU incentive regulation provides a safeguard for yearly adjustments of tariffs following for instance an X-factor or the regulatory asset value, in the US historically most of the agreed tariffs, once established, were not renegotiated.[34] Littlechild (2012) provided a detailed account of the process of negotiation of settlements through the FERC. He observed a trend towards increased settlement of rate cases by negotiation, instead of costly and time-consuming litigation. Yet day-to-day regulation has been reported to be "a complex process of exogenous regulation by FERC, self-regulation between pipeline and shippers, and market processes, e.g. for secondary capacity" (Von Hirschhausen 2008: 6).

When judging over a proposed business case, the FERC by law exclusively examines market manipulation and (a flexible variant of) efficiency, as "all rates and charges made, demanded, or received by any natural-gas company for or in connection with the transportation or sale of natural gas subject to the jurisdiction of the Commission, and all rules and regulations affecting or pertaining to such rates or charges, shall be just and reasonable."[35] Market manipulation is excluded by demanding third party access and fixing tariffs, when unreasonable or unjust prices are identified at any moment.[36] In November 2011, the standard rate of return that was allowed for new gas transmission pipelines was 14 percent. In comparison, the estimated WACC was 11.6 percent, based on data from 1996 to 2003 (Von Hirschhausen 2008: 7). This number according to FERC representatives functions as a market incentive, and hence seems to confirm earlier claims that the rate of return in the US is used as an instrument to attract investments in pipeline infrastructure (Joskow 2005). If operators and/or shippers consult the FERC in order to change the tariffs of an existing pipeline, due to operational costs, complaints of shippers, or other reasons, the regulatory authority can file a rate case and subsequently proposed a rate of return of 11.55 percent (in 2011). These pipeline rate cases are open to the public. It is worth noting that for the distribution grid the rates of return are lower, namely between 8 and 9 percent. In order to safeguard reasonable tariffs, FERC staff members advocated positions on behalf of the public interest in pipeline rate cases. In addition, the FERC has undertaken proceedings to reduce existing pipeline rates that it believes are no longer just and reasonable.[37]

The legal competences of the FERC focus on security of supply, as shown in Table 5.2. In the EU other policy goals are agreed upon and, as was shown, regulatory authorities can have them as core competences. Yet in the US these policy goals are not ventilated in US natural gas regulations.[38] In practice this means that the FERC operates in line with its philosophy that new pipeline facilities and expanding interstate pipeline grids increase the overall safety of the industry (allowing for older facilities to be abandoned) and hence enlarge reliability and efficiency.

Table 5.2 Focus of US regulatory authority FERC, in terms of efficiency, security of supply and sustainability

Country	Efficiency	Security of supply	Sustainability
United States	Delegated regulatory task in the US, though rate-of-return regulation is used as an investment vehicle and may subsequently prohibit most efficient transport tariffs	Delegated regulatory task for FERC that uses substantial rates of return for pipeline operators to invest	No delegated task for the FERC

Regulatory authorities in the EU take a fundamentally different approach towards calculating transport tariffs than their US federal counterparts. Regulatory authorities in both Great Britain and the Netherlands apply forms of incentive regulation and focus on efficient tariffs while sharing tasks and responsibilities regarding security of supply with legislature. On top of that, in Great Britain sustainability has been an explicit legal task of OFGEM since 2008. In the US, the FERC focuses exclusively on helping the market function, by applying rate of return regulation and only occasionally – though increasingly (Littlechild 2012) – renegotiating tariffs as set between system operators and shippers. The difference between allowed revenues is remarkable, with EU rates of return wobbling around 5 percent while investors in new transmission pipelines in the United States can count on almost triple that value.[39]

Private and public ownership, and investments

Theory does not provide a unanimous verdict regarding a preference towards public or private ownership of gas transmission companies. Viscusi *et al.* (2005: 508ff.) observed that although some studies report that regulated private electric utilities seem to perform more efficiently than publicly owned utilities, the evidence is not strong. In a broader analysis of infrastructure quality in deregulated industries, Buehler *et al.* (2004) concluded that under reasonable demand assumptions, investment incentives turn out to be smaller under vertical separation than under vertical integration. Regarding the gas industry, where liberalization and privatization generally means vertical unbundling or separation, this would be an argument against private ownership. Kwoka (2006: 146) argued that "while often suspected of inferior cost performance, the evidence here shows that publicly owned utilities achieve costs comparable to those under competition. As between those two regimes, public ownership appears more successful in controlling costs by itself, though regulation buttressed by benchmark competition achieves a similar result." Jamasb and Pollitt (2007: 6164), in a study of electricity markets in Great Britain concluded that "empirical evidence on the merits of private ownership and privatization in the context of market-oriented infrastructure reforms can be characterized as inconclusive. However, when accompanied by effective regulation, privatization has achieved efficiency improvements." Von Hirschhausen *et al.* (2004) nuanced the argument for privatization of infrastructure in terms of over-investment versus underinvestment. According to them, privatization in the 1980s was largely driven by the lack of public funds for infrastructure investment. In order to avoid underinvestment the responsibility was simply shifted to the private sector (*ibid.*: 209). De Joode (2012), in a study of regulation of gas infrastructure expansion, concluded that in particular the European case demonstrates that policy makers initially have failed to recognize the potential of competition and private capital in gas infrastructure investment. He argued that the regulatory framework should be geared to facilitate such investments in the future.

Cambini and Rondi (2010), in a study on the relationship between investment and regulatory regimes (incentive regulation versus rate of return), found that investments are higher under incentive regulation regimes. They also found that there is "no empirical evidence that private ownership boosts investment incentives" (*ibid.*: 4). This is remarkable, since theory suggests that incentive regulation carries the potential risk of under-investment: reduction of investments leads to higher return and can therefore be tempting. However, analysis has shown that in Dutch electricity and gas networks since 2001, incentive regulation has "ensured a more rational and professional approach towards investments, with investment levels coming down somewhat at the start of the regulation but picking up later on" (Haffner *et al.* 2010: 35).

While the academic jury on this topic is out, regulatory authorities are occupied with the question of how to generate sufficient interest to invest, albeit from private or public investors. It seems that the FERC has chosen the path of least resistance, by allowing significantly higher rates of return on gas transmission investments than European regulatory authorities do. In addition, with rate cases only being filed when users or operators bring the case to the regulatory authority's attention or when the regulatory authority itself decides to put a case to the test, investors have a reasonable period of certainty to get their money's worth. This leaves the question open whether the rates of return in the US trigger overinvestment and inefficient use of capital. It is difficult to conclude this based on the available evidence, though the FERC does not rule this out. One study on utilization rates of gas infrastructure concluded that evidence from the US suggests that lower utilization rates should in fact be linked to market development and integration: in a mature market, more gas infrastructure is needed not just to facilitate larger amounts of gas being consumed, but also to facilitate an increase in supply flows. Hence utilization rates may well drop, in accordance with US figures (Correljé *et al.* 2009: 17). More empirical work, for instance a closer examination of utilization rates of gas infrastructure facilities or end-user tariffs for gas consumption, would be helpful here, but is beyond the scope of this study.

With significantly lower rates of return, the British regulatory authority – like its European colleagues – has been occupied with the question how to attract additional investment. It boasts about its so-called regulatory asset value, which is the value upon which investors earn a return in accordance with the regulatory cost of capital. It is based on the historical investment costs and is set yearly, to complement existing longer term rates of return. In addition to the asset value, OFGEM in 2012 established new rates of return, for a longer regulatory period (2013–2021). The main reason for this three-year extension of the regulatory period is to provide more long-term security about the rates of return (i.e. to attract more long-

term capital intensive investors). As for the Dutch case, where the legislature gas been considering whether a minority share of the transmission
system operator could be privatized to attract more capital; the aforementioned experiences in Great Britain demonstrate that even when private
capital is involved, attracting investment is a complicated regulatory task. It
suggests that privatization of gas infrastructure cannot be considered as a
panacea. In Great Britain, extending the regulatory period may provide
additional stability for investors.

Decision-making structures, based on a multilevel governance analysis

Tables 5.3 and 5.4 explore what can be learned from decision-making structures regarding gas transmission infrastructure. Hereto a multilevel
governance framework is used. It is worth noting, that as multilevel governance theory was designed as a theory of European integration, a valid
comparison with the US case requires a shift upward in the scheme, treat-
ing the federal level in the US as the supranational level in the EU, the state
level in the US as the EU member state or national level, and so on.

The EU analysis requires several elucidations. First, the horizontal arrow
in Table 5.3 indicates that the clear distinction between the public and the
private domain is in fact too rigid. To give an example, the Dutch state
owns all the shares of the transmission system operator, Gas Transport
Services (GTS), as part of Gasunie. GTS is currently operating in the public
domain (i.e. regulated gas transmission activities). Yet the activities of
Gasunie are broader, with also investments in gas storage projects and LNG
facilities, in this case commercial activities that fit the private domain (a
situation that is not unique in Europe, as for instance also the Belgian TSO
Fluxys operates an LNG terminal and gas storage facilities in Belgium). In
addition, in the Netherlands there is an ongoing discussion about the privatization of a minority share of Gasunie, in order to attract additional
financial means. In other parts of the EU, such as Great Britain, this privatization of transmission networks has already taken place.[40]

Second, the vertical arrow indicates that institutions such as ACER and
its advisory body CEER operate on the supranational level. Subsequently, it
is worth mentioning, that though these supranational and national organizations in theory are complementary, this is debatable as far as their
working in practice is concerned. In principle their interests may conflict,
for increased decision-making power on one level of governance automatically implies a decrease of power on another level of governance. Regarding
the member state or national level the independent position of national
regulatory authorities is worth mentioning. Note that this situation is
comparable to that in the US, where FERC is an independent regulatory
agency. Contrary to the illustration from the Netherlands, in the US no legal
quarrels that can be linked to this status have been reported, but more

Table 5.3 European Union decision making regarding gas transmission infrastructure (based on the case of the Netherlands, taking other options in the European Union into consideration)

Governance level	Public domain	Private domain
Supranational/EU	Gasunie was the first European TSO to purchase an international network (2008). In December 2012, Belgian TSO Fluxys followed with the purchase of a 32% share in the Algerian/Spanish pipeline Medgaz. ACER has a limited mandate so far. Replaced ERGEG in 2011 (set up by EC to give advice on internal market).[a] ENTSO-G (TSO platform) and CEER promote completion of the internal gas market in Brussels but have no decision-making power (note that their individual members do, on the national level).	External investors in energy networks; an example is the role of Mitsubishi in Germany (as an investor in electricity networks)
National/EU member states	Gas transmission companies, differing per member state National governments drafting legislation and implementing EC guidelines and regulations National regulatory authorities safeguarding implementation of guidelines, fair competition in member states and controlling transmission tariffs with different regulatory systems Court of Appeal (CBb)	Gas transmission companies, differing per member state Investors in transmission pipelines Shippers Purchasers of natural gas Related industries, such as engineering firms, construction companies, compressor manufacturers, IT companies, banks, etc.
Regional	Joint decision making of regional authorities on environmental issues with regard to new pipelines (in some cases overrule from the national legislature is possible) Also differing per member state, more research needed	N/A
Local	See regional decision making	N/A

Note: [a] See page 16 and onward at http://eur-lex.europa.eu/LexUriServ/LexUriServ.do?uri= OJ:L:2011:129:FULL:EN:PDF (accessed September 12, 2012).

Table 5.4 United States decision making regarding gas transmission infrastructure

Governance level	Public domain	Private domain
Federal	FERC • Regulate interstate transmission pipelines • Review of investment proposals (incl. LNG terminals) and determine rate setting methods • Siting/abandonment of pipelines • Set rules for business practices • Take the lead on environmental reviews under environmental and preservation acts • Oversee mergers and acquisitions of pipelines, with Department of Justice, Federal Trade Commission, IRS and NRC US Environmental Protection Agency Assist on federal and state level in determining if environmental aspects of pipeline meet acceptable guidelines Department of Transport's Pipeline and Hazardous Materials Safety Administration, acting through Office of Pipeline Safety • Administers national regulatory program to assure safe transportation of natural gas, petroleum and other materials by pipeline • Certify safe operations, with jurisdiction over lifetime of pipeline Department of Energy (marginal role)	Investors in interstate pipelines Shippers Purchasers of natural gas Related industries, such as engineering firms, construction companies, compressor manufacturers, information technology service companies, banks, accountants and so on
State	Regulatory Utility Commissioners (also in National Association, i.e. NARUC) dealing with state regulations (e.g. pipeline safety Section Decision making regarding interstate pipelines takes place on federal level) Intrastate pipelines and distribution networks fall outside the scope of this study	Investors in intrastate pipelines. Note that gathering lines (from well head to compressor station) are not regulated Shippers Related industries, such as engineering firms, construction companies, compressor manufacturers, information technology service companies, banks, accountants and so on
Regional and local	N/A	N/A

research would be useful here. The analysis confirms earlier observations about the relations between legislatures and regulatory authorities. While European gas markets have been liberalized and up-scaled to the EU level, regulation is still a national affair. Following Thatcher (2005), on the EU level regulatory agency ACER surely is not a "distinct actor," contrary to national regulatory authorities. The consequences of the lack of EU regulatory orchestration deserve more empirical attention. Ruester *et al.* (2012a), despite the limited EU involvement, saw neither a need nor solid justification for an EU-wide harmonization of the regulation of TSO revenues. On the other hand, independent regulatory approaches differ per member state, and in combination with market players and increasingly also infrastructural companies that operate internationally, it may be expected that this mismatch can cause friction and does not contribute to the desired completion of the internal gas system. Also, in parts of the EU attracting investments in gas infrastructure continues to be cumbersome (as demonstrated for instance in European Commission 2014a). As is described in more detail in Chapter 7, in parts of the EU investments continue to hamper, and in particular cross-border projects are not being built. Ruester *et al.* (2012b) concluded that the heterogeneity in regulatory regimes in Europe probably does not hinder investments without cross-border impact, but that in particular in natural gas markets there is a clear distinction between national and international projects (see also European Commission 2014b). Ruester *et al.* (2012a) suggested considering innovative solutions such as competitive tendering or a European tariff component to collect funds from grid owners that can be reinvested in energy infrastructure.

With regard to decision-making structures in the US, several elements stand out. First, the division of labor is unequivocal, as shown in Table 5.4. There are clear roles for system operators, investors and federal regulatory authority. The legislature has adopted a laissez-faire approach, but can interfere when considered necessary, for instance through the Department of Transport. Decisions regarding interstate gas transmission infrastructure are made in the private sector, within boundaries indicated by organizations at the federal level. One caveat is that other types of infrastructure, in particular intrastate gas pipelines (that can also be gas transmission pipelines), are not part of this analysis. More research would be helpful here.

Discussion

At the beginning of the discussion of this chapter it is worth reiterating some of the fundamental differences between the US and EU gas systems, most notably the substantial difference in institutional history. While the US gas system developed during a decades-long period of regulatory reforms, the EU gas system is arguably in transition and may well have several decades of reforms ahead of it (as argued by Makholm 2012). Other substantial differences between the EU and US are discussed in detail in

Chapter 7. Despite these differences, this chapter aimed to analyze whether lessons can be learned from the US system, in particular regarding the incentives for investments in gas infrastructure, which are indispensable to make the internal EU gas system function.

Several conclusions follow from the analysis of transatlantic regulatory regimes regarding gas infrastructure. First, being a European frontrunner when it comes to liberalizing its energy markets, in Great Britain the regulatory authority explores the boundaries of its mandate (e.g. when it comes to the strict separation between legislature and regulatory authority, as required by European institutions). Examples of this are the explicit mandate of OFGEM to contribute to sustainability by means of its regulation or the decision to extend the regulatory period in Great Britain with three more years to provide private investors in infrastructure projects with more long-term stability.

The position of independent regulatory authorities deserves more empirical attention. Though the independence as safeguarded in Europe is comparable to the US, where the FERC acts independently from US legislatures, in Europe currently having independent national regulatory authorities means having 28 potentially different regulatory regimes. In a liberalized EU gas system a more coordinated European approach seems obvious, though there is some evidence to suggest that there is no need for full harmonization, because at the member state level investments do not seem to be hampered because of the observed heterogeneity of regulatory regimes. That is different, however, when it comes to cross-border investments, which are probably hindered (partly) because of this same heterogeneity. There are several instruments that could help trigger additional investments, though as observed and discussed in more detail in Chapter 7, these only seem to work in a part of the European gas system. From an academic point-of-view the observed changes in balance of power are worth further examination. While scholars have focused on the motives of politicians to create independent regulatory authorities and the consequences this has, with the creation of European regulatory agency ACER arguably a new tug-of-war has emerged within the EU gas system (i.e. with regard to the balance of power between this supranational regulatory agency and independent national regulatory authorities). This confirms an earlier analysis of Thatcher (2011) who concluded that formal delegation of mandate to European regulatory agencies has been limited and uneven. Several arguments against the position of these authorities have been described, such as the lack of accountability or the lack of democratic legitimacy (Christensen and Laegreid 2007; Maggetti 2009). As this case study showed, fundamental policy goals (i.e. sustainability, security of supply and efficiency) are likely to conflict at some stage. This was confirmed by for instance Szydlo (2012), who even concluded that the position of independent regulatory authorities is in conflict with essential constitutional principles such as the domain of the law.

Second, the lack of underinvestment in gas infrastructure in the US may be attributed to the substantial rates of return that private investors have been allowed to accrue, a mechanism that is assumed to take care of security of supply. In doing so, FERC uses US regulation as an investment vehicle to trigger investments in gas infrastructure. Yet it may well be that users of pipeline infrastructure and consumers in fact collectively pay too much for their gas transport. More research is needed to confirm this. Regardless, the analysis shows that EU regulatory authorities are occupied focusing on efficiency, and less on security of supply and sustainability, whereas in the US security of supply prevails and there are hints that efficiency is less important to its regulatory authority FERC. It is worth reiterating that in recent years in fact substantial gaps are observed between the increase in US natural gas production and investment levels in gas infrastructure, leading to bottlenecks in parts of the country, particularly in New England. At this point, however, there is no empirical evidence to support the notion that this is a structural problem of underinvestment, but rather an indicator of the significant pace at which domestic gas production continues to be ramped up.

Third, it is difficult to link the scope for underinvestment in gas infrastructure in Europe to either private or public capital. While the academic debate provides no final verdict, it is worth noting that the substantial investment levels as reported in the US cannot be linked to private investors per se. It seems plausible that the combination of private capital with rate of return regulation as it is applied in the US generates the level of investments in gas infrastructure that is currently witnessed. It is worth bearing in mind, however, that in the US system gas infrastructure projects that are not economic are simply not being built. Subsequently millions of American consumers are not connected to gas networks and have to apply alternative fuel sources, such as propane and wood. The case of Great Britain seems to indicate that attracting private capital in itself is no guaranteed panacea. Despite the fact that is has privatized its gas networks, according to OFGEM representatives the system still struggles to attract sufficient investment. As an example of this, OFGEM has extended the regulatory period with three years to provide potential investors with more long-term stability.

In terms of decision-making structures, the most fundamental difference between the EU and US is that the latter has clear mandates for different stakeholders and decision makers, while the opposite can be argued about the EU. Diverse institutional history and development is contributing to today's patchwork in Europe, with market players operating internationally and transmission system operators investing both nationally and internationally (an emerging phenomenon, though this is expected to increase). Although interesting from an academic point-of-view, this dynamic institutional development may be somewhat too dynamic for investors, which seek wealth maximization instead of asymmetric information, ineffective institutions, unpredictable rules and decisions, and unpredictable transaction costs, to speak with North (1991).

Currently the regulatory focus in the EU is at the member state level, which is not in line with other elements of the EU gas system. Whereas in Europe the traditionally US phenomenon of an independent regulatory authority has been adopted at the member state level, at the EU level regulatory agency ACER lacks formal decision-making powers and is clearly not a distinct actor like its US counterpart the FERC is. Though it seems that more "Europe" is needed to streamline EU regulatory regimes, this requires transfer of power from the member state level to the European level. Up to date national regulatory authorities appear reluctant to do so. Therefore it is worth considering that also substantially better coordination and consensus building between national regulatory authorities could be an improvement to the EU energy system as a whole. Though it may be difficult to imagine with the current patchwork of regulatory regimes in Europe, there are examples of international cooperation in which national authorities coordinate their efforts effectively (e.g. NATO). This illustrates that expanding ACER's mandate and shifting power to the supranational level is not a *conditio sine qua non* to make the EU energy system function better.

Acknowledgements

Special thanks go to Professor Machiel Mulder of the Dutch regulatory authority Nederlandse Mededingings Autoriteit (NMa), and Brian S. White and Michael Strzelecki of the United States Federal Energy Regulatory Commission (FERC), who have helped with earlier versions of this chapter. I would also like to thank my former colleagues of the Transatlantic Academy and those of the German Marshall Fund who have given useful comments and feedback on that first version, which was published as a policy paper under the auspices of the German Marshall Fund of the United States.

Notes

1   *Ibid.*
2   Note that some of the other fundamental differences between the US and EU gas system are discussed in Chapter 7.
3   Most academic literature on regulation focuses on electricity and not on natural gas. According to Kwoka and Madjarov (2007: 27) this has to do with special characteristics of electricity and the dynamic that this brings for regulation (e.g. non-storability, low demand elasticity, and high capital intensity).
4   Gaswet, article 10a, sub 1, part a. Note that indirectly security of supply does play a role in the Dutch case, as the analysis shows later.
5   In 2005 the integrated gas company Gasunie was unbundled into the commercial company GasTerra and the public transmission system operator GTS. The gas transport facilities that are not part of the national transmission system are administered by regional infrastructure companies. In the Netherlands there are

twelve of these companies, namely Cogas, Delta, Enexis, Liander, NRE, Rendo, Stedin, Westland, Haarlemmermeer, Obragas, Intergas, and Zebra.

6   Represented by the Treasury.

7   See (in Dutch) FIN/2012/969M and the report of American Appraisal (2012) about this purchase by Gasunie.

8   As announced on February 2, 2009. This is supported by an official press release of Dutch Gasunie on March 23, that stated it had no interest in Thyssengas ("kein weiteres Interesse") since the German investment climate had become unpredictable through regulatory changes ("Regulierung hat das Investitionsklima in Deutschland "unberechenbar" gemacht"). By late 2012 research that was carried out on behalf of the Dutch Finance Ministry concluded that Gasunie management made serious errors in the purchasing process and that the company paid too much for the German gas network (American Appraisal, 2012).

9   See for more details Commission staff working document SEC (2011) 755 final, p. 6ff.

10   As laid down in for instance the Lisbon Treaty, 2007/C, 306/01, article 176 A.

11   Report of the EP following the proposed regulation 1/2003 A5-0229/2001, June 21 2001, p. 22. See also Staatsblad 327, June 30, 2005.

12   To be more precise, in 2006 CBb ruled that regulatory format and existing legal framework did not fit. The then Minister of Economic Affairs Van der Hoeven issued a new policy rule, among other setting concrete parameters regarding GTS's capital expenditures. CBb, however, in 2010 argued that the Minister had impinged on the independent decision-making process of the Dutch regulator NMa. See also www.nma.nl/en/documents_and_publications/press_releases/ news/2011/11_21_nma_makes_draft_method_decisions_gts_available_for_per usal.aspx.

13   LJN: BM9470, June 29, 2010. Comparable verdicts were published on the regulations for electricity TSOs and DSOs in that same period.

14   See www.nma.nl/en/documents_and_publications/press_releases/news/2011/ 11_49_nma__dutch_gas_transmission_system_operator_is_to_return_eur_400 _million_to_its_customers.aspx. For a report of the final verdict from November 2012, see www.energiekeuze.nl/nieuws.aspx?id=1354.

15   EC staff working document SEC (2011) 755 final, p. 5.

16   For an overview of major regulatory changes, visit www.eia.gov/oil_gas/ natural_gas/analysis_publications/ngmajorleg/ngmajorleg.html.

17   See also www.aga.org/Kc/aboutnaturalgas/Pages/default.aspx.

18   These data are derived from interviews with representatives of the FERC and the American Gas Association. More empirical work into the competitive model of gas infrastructure in the US and its consequences would be helpful.

19   See www.ingaa.org/File.aspx?id=21498.

20   Directive 2003/55/EC, art. 25, sub 1.

21   *Ibid.*, article 25, sub 2.

22   *Ibid.*, article 5.

23   Directive 2009/28/EC, article 16.

24   *Ibid.*, article 16, sub 2(c).

25   Quoted from website http://energy.gov/mission (accessed November 25, 2011).

26   Note that the FERC has no jurisdiction at the distribution level, following Natural Gas Act article 717, sub b and c. State level regulation may apply to distribution networks, but that does not fall within the scope of this book.

27   Docket no. RM91-11-002. Order no. 636-A section II.

28   See National Grid investor update from 2006: www.nationalgrid.com/NR/ rdonlyres/0C5D350E-3B87-469B-BD80-73895876C953/13654/NGTPCR4 overview15DEC06FINAL.pdf.

29   Energy Act of 2008, part 5, art. 83, sub 1c.
30   For more information, see chapter 11 of www.ofgem.gov.uk/Networks/Trans/Archive/TPCR4/ConsultationDecisionsResponses/Documents1/16342-20061201_TPCR%20Final%20Proposals_in_v71%206%20Final.pdf.
31   For a recent example, see www.ofgem.gov.uk/Sustainability/SDR/Documents1/Sustainable%20development%20focus%202011_WEB.pdf.
32   Gaswet, article 12, sub 1.
33   See www.nma.nl/images/Bijlage%202%20WACC%20bij%20Methodebesluit%20Transport%20GTS%202010-2013%20(2)22-193277.pdf.
34   As confirmed in an interview with representatives of the FERC on November 23, 2011.
35   Natural Gas Act, section 5, art. 717C, sub a.
36   *Ibid.*, art. 717C-1 and art. 717D, sub a.
37   Data derived from interviews with FERC representatives in 2011 and 2012.
38   It is worth noting that the FERC does share strategic goals that resemble the ones defined in the Lisbon Treaty, such as the Strategic Plan 2009–2014 states that FERC's mission is to "assist consumers in obtaining reliable, efficient and sustainable energy services at a reasonable cost through appropriate regulatory and market means"; information derived from FERC's strategic plan, consulted online on 23 November at www.ferc.gov/about/strat-docs/FY-09-14-strat-plan-print.pdf.
39   Regarding distribution infrastructure the rate of return is lower, but still twice as large in the US as compared to the EU.
40   National Grid plc operates gas transmission and distribution networks in the UK and in northeastern US. The company was first listed on the London Stock Exchange in 1995.

References

American Appraisal, 2012. *Onderzoeksrapport: Overname Gasunie Deutschland Transport Services GmbH (voorheen geheten: BEB Transport GmbH) door N.V. Nederlandse Gasunie.* Rotterdam: American Appraisal.

Ascari, S., 2011. *An American Model for the EU Gas Market?* Working paper RSCAS 2011/39. San Domenico di Fiesole: Robert Schuman Center for Advanced Studies, European University Institute.

Asche, F., Misund, B., Sikveland, M., 2013. The relationship between spot and contract gas prices in Europe. *Energy Economics* 38; 212–217.

Boersma, T., Ebinger, C. K., 2014. *Prevailing Debates Related to Natural Gas Infrastructure.* Natural Gas Briefing Document 3. Washington, DC: The Brookings Institution.

Buehler, S., Schmutzler, A., Benz, M.-A., 2004. Infrastructure quality in deregulated industries: is there an underinvestment problem? *International Journal of Industrial Organization* 22(2): 253–267.

Cambini, C., Rondi, L., 2010. Incentive regulation and investment: evidence from European energy utilities. *Journal of Regulatory Economics* 38(1): 1–26.

Christensen, T., Laegreid, P., 2007. Regulatory agencies: the challenges of balancing agency autonomy and political control. *Governance: An International Journal of Policy, Administration and Institutions* 20(3): 499–520.

Corr#ljé, A., De Jong, D., De Jong, J., 2009. *Crossing Borders in European Gas Networks: The Missing Links.* The Hague: Clingendael International Energy Programme.

De Joode, J., 2012. *Regulation of Gas Infrastructure Expansion*. Delft: Next Generation Infrastructures Foundation, Delft University of Technology.

De Vany, A., Walls, D., 1994. Natural gas industry transformation, competitive institutions and the role of regulation. *Energy Policy* 22(2): 755–763.

Elgie, R., 2006. Why do governments delegate authority to quasi-autonomous agencies? The case of independent administrative authorities in France. *Governance: An International Journal of Policy, Administration, and Institutions* 19(2): 207–227.

European Commission, 2011. *Proposal for a Regulation on Guidelines for Trans-European Energy Infrastructure and Repealing Decision Number 1364/2006/EC*. COM(2011) 658 final. Brussels: European Commission. See http://eur-lex.europa.eu/LexUriServ/LexUriServ.do?uri=COM:2011:0658:FIN: EN:PDF.

European Commission, 2012. *Communication: Making the Internal Energy Market Work*. COM(2012) 663 final. Brussels: European Commission.

European Commission, 2014a. *Communication on the Short Term Resilience of the European Gas System*. COM(2014) 654 final. Brussels: European Commission.

European Commission, 2014b. *Communication: European Energy Security Strategy*. COM(2014) 330 final. Brussels: European Commission.

Gasmi, F., Oviedo, J. D., 2010. Investment in transport infrastructure, regulation, and gas-to-gas competition. *Energy Economics* 32(3): 726–736.

Haffner, R., Helmer, D., Van Til, H., 2010. Investment and regulation: the Dutch experience. *Electricity Journal* 23(5): 34–46.

Heather, P., 2012. *Continental European Gas Hubs: Are They Fit for Purpose?* Oxford: Oxford Institute for Energy Studies.

Herbert, J. H., Kreil, E., 1996. Viewpoint: US natural gas markets – how efficient are they? *Energy Policy* 24(1): 1–5.

INGAA, 2014. *North American Midstream Infrastructure through 2035: Capitalizing on Our Energy Abundance*. See www.ingaa.org/File.aspx?id=21498.

Jamasb, T., Pollitt, M., 2007. Incentive regulation of electricity distribution networks: lessons of experience from Britain. *Energy Policy* 35(12): 6163–6187.

Jamasb, T., Pollitt, M., Triebs, T., 2008. Productivity and efficiency of US gas transmission companies: a European regulatory perspective. *Energy Policy* 36(9): 3398–3412.

Johnson, C., Boersma, T., 2013. Energy (in)security in Poland, the case of shale gas. *Energy Policy* 53: 389–399.

Joskow, L., 2005. Supply security in competitive electricity and natural gas markets. See http://economics.mit.edu/files/1183 (accessed September 11, 2012).

Kwoka, J., 2006. The role of competition in natural monopoly: costs, public ownership, and regulation. *Review of Industrial Organization* 29(1–2): 127–147.

Kwoka, J., Madjarov, K., 2007. Making markets work: the special case of electricity. *Electricity Journal* 20(9): 24–36.

Larsen, A., Pedersen, L. H., Sørensen, E. M., Olsen, O. J., 2006. Independent regulatory authorities in European electricity markets. *Energy Policy* 34: 2858–2870.

Lévêque, F., Glachant, J.-M., Barquín, J., Van Hirschhausen, C., Holz, F., Nuttall, W. J., 2010. *Security of Energy Supply in Europe: Natural Gas, Nuclear and Hydrogen*. Northampton, MA: Edward Elgar.

Littlechild, S., 2012. The process of negotiating settlements at FERC. *Energy Policy* 50: 174–191.

Maggetti, M., 2009. The role of independent regulatory agencies in policy-making: a comparative analysis. *Journal of European Public Policy* 16(3): 450–470.

Makholm, J. D., 2012. *The Political Economy of Pipelines: A Century of Comparative Institutional Development*. Chicago, IL: University of Chicago Press.

Medlock III, K. B., 2012. Modeling the implications of expanded US shale gas production. *Energy Strategy Reviews* 1(1): 33–41.

Mohammadi, H., 2011. Market integration and price transmission in the US natural gas market: From the wellhead to end user markets. *Energy Economics* 33(2): 227–235.

Murry, D., Zhu, Z., 2008. Asymmetric price responses, market integration and market power: a study of the US natural gas market. *Energy Economics* 30(3): 748–765.

North, D. C., 1991. Institutions. *Journal of Economic Perspectives* 5(1): 97–112.

Pelletier, C., Wortmann, J. C., 2009. A risk analysis for gas transport network planning expansion under regulatory uncertainty in Western Europe. *Energy Policy* 37(2): 721–732.

Renou-Maissant, P., 2012. Toward the integration of European natural gas markets: a time-varying approach. *Energy Policy* 51: 779–790.

Ruester, S., Marcantonini, C., He, X., Egerer, J., Hirschhausen von, C., Glachant, J.-M., 2012a. *EU Involvement in Electricity and Natural Gas Transmission Grid Tarification*. Policy brief 2012/01. Florence: Florence School of Regulation.

Ruester, S., von Hirschhausen, C., Marcantonini, C., He, X., Egerer, J., Glachant, J.-M., 2012b. *EU Involvement in Electricity and Natural Gas Transmission Grid Tarification*. Final report. Florence: RSCAS, European University Institute.

Siliverstovs, B., L'Hégaret, G., Neumann, A., von Hirschhausen, C., 2005. International market integration for natural gas? A cointegration analysis of prices in Europe, North America and Japan. *Energy Economics* 27(4): 603–615.

Szydlo, M., 2012. Independent discretion or democratic legitimization? The relations between national regulatory authorities and national parliaments under EU regulatory framework for network-bound sectors. *European Law Journal* 18(6): 793–820.

Ter-Martirosyan, A., Kwoka, J., 2010. Incentive regulation, service quality, and standards in US electricity distribution. *Journal of Regulatory Economics* 38(3): 258–273.

Thatcher, M., 2002a. Delegation to independent regulatory agencies: pressures, functions and contextual mediation. *West European Politics* 25(1): 125–147.

Thatcher, M., 2002b. Regulation after delegation: independent regulatory agencies in Europe. *Journal of European Public Policy* 9(6): 954–972.

Thatcher, M., 2005. The third force? Independent regulatory agencies and elected politicians in Europe. *Governance: An International Journal of Policy, Administration, and Institutions* 18(3): 347–373.

Thatcher, M., 2011. The creation of European regulatory agencies and its limits: a comparative analysis of European delegation. *Journal of European Public Policy* 18(6): 790–809.

Vazquez, M., Hallack, M., Glachant, J.-M., 2012. *Building Gas Markets: US versus EU, Market versus Market Model*. Working paper RSCAS 2012/10. San Domenico di Fiesole: Robert Schuman Center for Advanced Studies, European University Institute.

Viscusi, W. K., Harrington Jr., J. E., Vernon, J. M., 2005. *Economics of Regulation and Antitrust*, fourth edition. Cambridge, MA: MIT Press.

Von Hirschhausen, C., 2008. Infrastructure, regulation, investment and security of supply: a case study of the restructured US natural gas market. *Utilities Policy* 16(1): 1–10.

Von Hirschhausen, C., Beckers, T., Brenck, A., 2004. Infrastructure regulation and investment for the long-term: an introduction. *Utilities Policy* 12(4): 203–210.

6 The political minefield called European shale gas

Introduction

Encouraged by developments that have taken place since roughly 15 years in the US and motivated by a prospect that is labeled "energy security," Polish government officials, business representatives, and other specialists have been advocating the extraction of natural gas from shale rock layers under their soil. The Polish, together with the government of the United Kingdom, are the front-runners in the European shale gas debate. Despite many unanswered questions, the currently preferred technology of shale gas extraction, called hydraulic fracturing ("fracking"), has been embraced, and according to some is going to end dependence on Russian gas once and for all. Yet unlike the US, it is unclear whether in the EU one single molecule of shale gas is going to be produced.

This case study examines the shale gas extraction potential in the EU, while using the US as a benchmark. It does so by studying available evidence from the US, the only country worldwide where – at the time of writing – extraction of shale gas is full-fledged in process. It has uncovered a broad range of new challenges and concerns, which are related to market structure and functioning, the environment, infrastructure and regulation. These issues are addressed in the next sections. Furthermore the concept of energy security, which is frequently used in shale gas debates on both sides of the Atlantic Ocean, is discussed. It is worth noting that this part of the chapter may contain some overlap with the first part of Chapter 2, in which energy security studies were discussed. The case study ends with an institutional overview of decision-making structures, using a multilevel governance framework. The framework is slightly different from the one that was used in Chapter 5. While both in the US and the EU water related decision-making institutions can operate at an interstate level, this category was added to the original framework. When examining the viability of shale gas extraction in the EU the focus is primarily on Poland, since that country is currently most energetically striving for shale gas extraction. It is also the country where to date most exploratory wells have been drilled, while others have either banned available technologies (i.e. France and Bulgaria) or requested more time to evaluate environmental concerns (e.g. Germany, Netherlands) or adjust regulatory frameworks (e.g. Czech Republic). Other European member states will, however, be touched upon throughout the case study, whenever available evidence validates doing so.

The data in this chapter have been derived from the available academic literature, policy papers and reports and from interviews with business representatives and policy makers during fieldwork in Pennsylvania, Poland, and Brussels. The data cover the period up to December 2014.

Market developments

United States

Over the last decade shale gas production in the US has exploded. In 2007 roughly 1.3 trillion cubic feet (tcf) was produced (approximately 37 bcm), a number that rose to over 5 tcf (141.5 bcm) in 2010 and projections are that this number will almost triple by 2035 (US Energy Information Administration 2012a). However, the limitations of these projections have to be considered.

First, there can be new technological or geological insights. It is worth noting that little over one decade ago most people working on natural gas in industry, government and academia had not heard about shale gas at all. Hence, predicting future developments is exceedingly difficult. To illustrate this, the 2012 Energy Information Administration (EIA) Annual Energy Outlook substantially downgraded the estimates of technically recoverable reserves of shale gas for the US, largely due to a decrease in the estimate for the so-called Marcellus shale, from 410 tcf to 141 tcf (or 11.603 bcm to 4.000 bcm). Blohm *et al.* (2012) linked this substantial difference to existing reserve estimation techniques, which ignore current land use patterns, regulations and policies and therefore do not accurately represent the accessible reserves. The EIA reported that existing uncertainties about recoverable reserves can be linked to the relatively limited number of available data.[1] It is worth noting that large parts of the Marcellus shale lie in New York, where hydraulic fracturing is banned.[2] Even in Pennsylvania 85 percent of the wells are drilled in geographically concentrated areas, making estimates about recoverable reserves less reliable.[3] Also, most of these wells have only been drilled recently, making estimates about long-term production rates uncertain. These factors all contribute to the uncertainties regarding recoverable reserves. Yet despite these uncertainties it seems that the estimates as published annually by the EIA have typically been rather modest in retrospect. In 2014 the EIA estimated that natural gas production in the US continues to grow steadily, with a 56 percent rise between 2012 and 2040. Currently, it is believed that prices for natural gas stay comparatively low, which boosts energy intensive industries in the coming decades. It is also an incentive for an increased share of natural gas in electricity generation, eventually taking over coal as the dominant fuel source in the US around 2040 (US Energy Information Administration 2014).

Second, economic conditions of the gas market can change. From July 2008 natural gas wellhead (wholesale price at its point of production) prices have plummeted due to overproduction (US Energy Information Administration 2012b). Kaiser (2012) showed that in the Haynesville shale gas play, Louisiana, average wells cost between US$7 million and US$10 million to drill and complete, and break-even costs range between US$4 and US$6 per million cubic feet (mcf) of natural gas. That means that at prevailing gas prices most of the shale gas wells in this case study fail to break even. When gas prices fall below US$4 per mcf, only a small fraction of wells in the Haynesville play would be profitable. This development urged some to suggest the reintroduction of a wellhead price-floor, which was abolished in 1989 with the Wellhead Decontrol Act (Weijermars 2011). However, it appears that the market itself has corrected the existing mismatch between supply and demand, with for instance British Petroleum reporting a write-off of US$2.1 billion on shale gas acreage because of lower natural gas prices.[4] Also, since its peak in late 2008 the number of rigs used for shale gas extraction has been in decline, since low prices forced producers to look for alternative business (e.g. tight oil).[5] That does not mean that natural gas production is in decline though, as the share of associated natural gas (natural gas produced as a byproduct of for instance tight oil or gas liquids) has increased substantially, and according to the IEA will hover around 54 percent in 2014.[6] Thus, the EIA forecasts suggest that domestic wellhead prices shall remain below US$5 per thousand cubic feet until at least 2023. It is important to note that as technology improves, production costs continue to come down. Aguilera (2014) showed that improvements in technologies have reduced the costs of producing unconventional natural gas to the point where, in some cases, these costs are lower than conventional natural gas production. This in turn is expected to trigger investment in gas-fired electricity plants, leading to an assumed minimal share of 27 percent of natural gas in electricity generation by 2035 (Paltsev *et al.* 2011). More broadly speaking, the future of natural gas depends on a complex set of factors, such as adoption of natural gas for transportation, future climate policies, renewable energy policies or the lack thereof, and geopolitical considerations (Myers Jaffe and O'Sullivan 2012).

An open question that is relevant to US market development and domestic production is whether or not the US will export substantial amounts of its natural gas in the future. Current predictions are that the country can be a net exporter of natural gas by 2016 (US Energy Information Administration 2011). So far two companies have received an unrestricted license to export natural gas in the form of LNG, while others are still in the cumbersome licensing process.[7] Under the Natural Gas Act, the US Department of Energy must approve permit applications to export natural gas to the 15 countries that have free trade agreements (FTA's) with the US covering natural gas.[8] However, of these countries only Canada, Chile, Dominican Republic and Mexico have existing LNG terminals (Ratner *et al.* 2011). In 2012 Korea joined this group of FTA countries.[9] In the near term, exports by pipeline are realized, and expectations are that exports to Mexico in particular will grow with about 6 percent per year (US Energy Information Administration 2014). That raises new questions about the possible effect of these exports on Mexican domestic market reforms, but this is outside the scope of this book. Though possible amounts of LNG exports from the US have been fiercely debated and depend on a complex set of factors (domestic production, production elsewhere, demand, competition from other fuel sources, carbon policies, financing, etc.), it is safe to say that in the coming years the US will become a substantial exporter of LNG (for an overview of different estimates of future US LNG exports, see for instance Houser and Mohan 2014). It is highly uncertain, however, what volumes of US LNG will eventually reach the market. It is expected that the first shipments of LNG from the US will get into the market in late 2015, and time will tell how many companies will follow suit.

Opponents of LNG exports argue that exports may drive up domestic gas prices. Some first studies indeed indicated that exporting LNG can drive up domestic prices for natural gas (US Energy Information Administration 2012c). Others dispute this conclusion, stating that apparent profitable export options are based on current, but in fact transitory market conditions, that erode due to supply responses abroad (Medlock 2012a). In December 2012 the US Department of Energy published a study on the macro-economic effects of exports of natural gas from the US, which outcomes supported loosening existing restrictions on LNG exports (Montgomery *et al.* 2012). Following this report several US politicians initiated new legislation in January 2013, which would allow for LNG exports to certain countries, for instance NATO members.[10] It is worth noting that the US has been exporting natural gas since at least the 1930s – before that data were not collected – to Canada and Mexico and that these exports have risen with over 10 times since 1999 (Ratner *et al.* 2011). Except for electricity-fired power plants, however, no significant increase in demand for natural gas is expected in the US in the coming years (*ibid.*: 17).

Though exporting excessively produced natural gas may be an obvious solution, some studies have suggested that in a more integrated global gas market, much of the US shale gas is expected to be too costly to compete in Europe with conventional resources from the Middle East and Russia. In fact, a more integrated global gas market could result in significant US gas imports (Paltsev *et al.* 2011). Boersma *et al.* (2014) concluded that US LNG would probably be competitive in the most liquid markets in northwestern Europe (e.g. the United Kingdom and the Netherlands), but not in the larger part of continental Europe. Some medium-term forecasts suggested that a more integrated world gas market can be a long term scenario, for during the last years regional gas prices have been drifting further apart and divergent prices are expected to remain a feature of global gas markets (International Energy Agency 2012). Comparable to the future of natural gas in general, the future of LNG exports from the US depends on a complex set of factors, such as longer term shale gas developments outside the US, development of pipelines from Russia and central Asia to potential US export markets, the effect of exchange rate movements on dollar-denominated supplies, and the extent of liquidity in the market and consequences of moving away from oil-indexation of gas prices (Medlock 2012a). Therefore, it remains to be seen whether LNG trade in time leads to a global gas market similar to that for oil, as predicted by Deutsch (2011). Stern (2014) argued that pricing regimes for natural gas are different throughout Europe and that also in Asia a transition away from long-term contracts and some form of oil indexation is not expected any time soon.

Next to impacts on the US market for natural gas, effects of shale gas extraction on the ground are worth mentioning, for they often are an important argument used by shale gas development proponents. Clearly in states such as Pennsylvania, Texas and Oklahoma, once forgotten towns are blossoming again, in terms of new roads being constructed to facilitate intensive truck usage to accommodate water supply delivery, new hotels being built to house the workforce and increased revenues for local retailers. Yet it has been difficult to quantify these economic benefits, and estimates therefore vary widely. Rabe and Borick (2011) showed that Pennsylvanians have significant doubts about the credibility of the media, environmental groups and scientists on this issue, which follows from a survey of over 500 inhabitants. However, these people in majority believe that natural gas drilling has provided more benefits than problems and that this trend will continue in the future. Kelsey *et al.* (2011) have suggested that the initial assessments about economic benefits of shale gas extraction were too optimistic, indicating that not only were the benefits lower than expected, but also that only half of the revenues stays in the hands of the locals. Up to now in fact job creation has been modest compared to earlier estimates, and roughly 40 percent of the workforce has been reported to be non-resident (*ibid.*). This is not surprising given the highly specialized knowledge that is required, in particular in early phases of shale gas exploration and extraction. Empirical evidence from Colorado, Texas, and Wyoming suggested that earlier predictions about job creation may have been too large and that large increases in value of gas production caused modest increases in employment, wage and salary income, and median household income (Weber 2012). Even though initial estimates may have been too optimistic, by no means should the economic impacts of oil and gas extraction be underestimated. Houser and Mohan (2014) concluded that this phenomenon could probably not have come at a better time. While the overall US economy has been struggling in recent years due to the global recession, their research suggested that shale gas and tight oil production in the US could increase annual GDP growth by as much as 0.2 percent on average between 2013 and 2020, boosting economic output by a cumulative 2.1 percent until the end of the decade. In the long run that number is expected to decrease to 1.4 percent, as other sectors in the economy recover and will compete with the oil and gas industry for labor and capital (*ibid.*: 143). In terms of overall employment, Houser and Mohan observed that local impacts can be very significant, with North Dakota being the most prominent example of a state where unemployment figures have dropped dramatically. In other states like Pennsylvania the impact of the oil and gas industry is more modest.

European Union

Up to now, the effects of US shale gas production on European markets are exclusively indirect and this is not expected to change before the end of 2015. LNG from Qatar, other parts of the Middle East and also eastern Siberia intended for terminals in North America is now finding its way to European and, predominantly, Asian markets (as described in more detail in Chapter 7). Furthermore the increased usage of gas-fired power plants in the US due to record low gas prices has made coal cheap and available, and in combination with a dysfunctional carbon emission trading scheme in Europe this has resulted in an increase of coal-fired electricity generation in Europe (Rühl 2012). LNG is changing the dynamics of global gas markets and European gas prices on spot markets have been significantly lower than oil-indexed gas during recent years. It is therefore expected that the EU continues to move slowly away from oil-indexation (Pearson *et al.* 2012; Stern 2014).

Currently it is too early to tell whether domestic European shale gas production takes off and if it does, whether this gas can compete with cheap Russian, Norwegian, Algerian, or Dutch gas that is abundantly available on the market. A reconstruction of why the shale gas boom happened in the US shows that the interplay of favorable geological conditions, access to and availability of infrastructure, substantial public support, availability of service industries, broad political support, a large consumer market and a favorable fiscal climate created a unique momentum that is unlikely to be copied on the European continent (Boersma and Johnson 2013).

Significant shale gas resources have been reported in the EU (Leteurtrois *et al.* 2011; Polish Geological Institute 2012; US Energy Information Administration 2011). Yet given the absence of experience with shale gas extraction in most parts of the world and given the number of affiliated uncertainties, reserve estimates should be treated with "considerable caution" (Pearson *et al.* 2012). In contrast to the US, actual shale gas extraction is still in the embryonic phase. A replication of the US shale gas revolution has been questioned, with reference to less favorable geological conditions, the absence of tax breaks, lack of a well-developed onshore service industry and the possible lack of public support due to the absence of local financial benefits (Stevens 2010). The Joint Research Center predicted that in the long run the best case shale gas production scenario for the EU is replacement of declining conventional production and having import dependence not exceed 60 percent (Pearson *et al.* 2012).

In Poland, several handfuls of wells have been drilled in the last few years.[11] In comparison, in Pennsylvania alone in 2014 between January 1 and November 24 according to the Department of Environmental Protection a total of 1.986 new wells were drilled in the state.[12] Companies in Poland are examining cores to establish the quality of gas and calculate at what costs it can eventually be extracted. So far the results have been disappointing, with ExxonMobil ending its exploratory operations in Poland in June 2012 after two disappointing wells being drilled.[13] Later on, Talisman Energy, Marathon Oil, ENI, and 3 Legs Resources also left the country, due to bad geology but also bad governance.[14] Countries like Germany and the Netherlands are awaiting further research, particularly on environmental concerns that have been linked to shale gas extraction, and public opposition. France and Bulgaria have put outright bans on hydraulic fracturing, the currently preferred technology to extract natural gas from shale rock layers. According to French officials hydraulic fracturing brings too many uncertainties and moreover local benefits are too meager (Leteurtrois *et al.* 2011). Bulgarian authorities in January 2012 were even so enthusiastic to put a ban on shale gas extraction that they made low-pressure hydraulic fracturing for conventional drilling impossible in the process, an unintended consequence that was abolished in June 2012.[15] Czech Republic officials argued in the fall of 2012 that their current regulatory framework is not geared to safeguard shale gas extraction in an environmentally viable fashion and were therefore considering a ban on shale gas explorations until June 2014.[16]

Environmental concerns

This section provides an overview of the most prominent environmental concerns that have been linked to shale gas extraction and hydraulic fracturing.

Carbon footprint and fugitive methane

Gas that is released during the production process ("fugitive methane") can have implications for the atmosphere and groundwater. This section focuses on potential climate impacts of methane emissions and fuel switching to natural gas, whereas the subsequent section will discuss potential ramifications of methane as a groundwater pollutant.

Fugitive methane represents the gas that is leaked during the entire life cycle (i.e. from extraction to burning). As a greenhouse gas, methane is roughly 20 times more potent than CO_2 but it has a much shorter life cycle in the atmosphere (Alvarez *et al.* 2012). Methane emissions from natural gas production have increased by 25.8 teragrams of CO_2 equivalent, or 13.6 percent, since 1990 (US Environmental Protection Agency 2012a). Of the fugitive methane emissions 58 percent occur during field production (e.g. leakage from the wells, gathering pipelines or gas treatment facilities). In short, during the production of both conventional and unconventional natural gas methane leakages occur, and most of the academic debate focuses on the question how large these emissions are as a percentage of overall production.

Some studies have suggested that shale gas wells have substantially higher fugitive methane emission rates (between 3.6 percent and 7.9 percent) than conventional gas wells (between 1.7 percent and 6.0 percent) (Howarth *et al.* 2011). Other academics have questioned the data used in that paper (Cathles *et al.* 2012). Some have argued that technical fixes are available to substantially reduce the amount of fugitive methane (Wang *et al.* 2011; Jenner and Lamadrid 2013). Yet others have indicated that these fixes focus primarily on preproduction emissions, while life cycle estimates are mostly dominated by the combustion emissions of the gas (Jiang *et al.* 2011). Based on a nearly 4,000 shale gas well sample from 2010, O'Sullivan and Paltsev (2012) concluded that hydraulic fracturing operations have not substantially altered greenhouse gas emissions from the natural gas sector. According to their estimates based on "current field practice" 70 percent of potential fugitive emissions are captured using green completion technologies, while 15 percent of those potential emissions are vented and 15 percent is flared. Heath *et al.* (2014) concluded that based on harmonization of median estimates of GHG emissions from shale-gas generated electricity the emissions were similar to those of conventional natural gas, with both approximately half that of the central tendency of coal.

It is less certain what the impacts of increased usage of natural gas in the long term could be. There is substantial evidence that cheap natural gas has incentivized fuel-switching in the US electricity sector (i.e. substituting coal-fired electricity generation for gas-fired power; e.g. Newell and Raimi 2014). Their study, however, also suggested that in the long term cheap natural gas also competes with other fuel sources than coal (e.g. renewables and nuclear), and that therefore there may be a limited long-term impact of increased natural gas use on carbon emissions (see also McJeon *et al.* 2014). Also, increased natural gas production is expected to result in lower natural gas prices, thus serving as a boon for additional consumption. Thus, Newell and Raimi (2014) concluded that increased natural gas production slightly increases energy use, substantial fuel-switching and that there is likely a slight alteration in terms of economy wide GHG emissions, depending on modeling assumptions. One of the assumptions that is fiercely debated, are upstream emissions (as discussed before) and also leakage of methane from natural gas infrastructure. Brandt *et al.* (2014) concluded that although official estimates consistently underestimate actual MH_4 emissions, system-wide leakage is unlikely to be large enough to negate climate benefits that come with coal to natural gas substitution.

Thus, uncertainties remain in this debate, whether shale gas is potentially a viable bridging fuel to a low carbon economy and what can in fact be done about methane emissions during the life cycle. These uncertainties led Stephenson *et al.* (2012) to the conclusion that the frequently used terminology of natural gas being an ideal "transition fuel" to a low carbon economy should be abandoned. Myhrvold and Caldeira (2012) have concluded that large scale usage of natural gas is not the way forward in the transition to low-carbon electricity, and have suggested a combination of conservation, wind, solar, nuclear energy and possible carbon capture and storage instead. Ironically, some observers have noted that carbon sequestration sites could be restricted due to large-scale shale gas extraction from shale rock layers. In short, shale gas extraction involves the fracturing of shale rock layers in order to increase its permeability to let the natural gas flow up into the well and as such hydraulic fracturing is in conflict with using these rock formations as a barrier to CO_2 migration (Elliot and Celia 2012). Newell and Raimi (2014) concluded that natural gas helps reduce GHG emissions, but that targeted climate policy measures are required to significantly drive down economy-wide emissions. They also noted that cheap natural gas may help to bring down the costs of installing those types of policy.

Contamination of ground water and surface water

Next to air quality there also continue to be concerns about shale gas extraction and its consequences in terms of water quality. So far, two cases of contaminated drinking water were reported that were likely directly linked to shale gas extraction. The first is in Pavillion, Wyoming, where the Environmental Protection Agency (EPA) started investigating private water wells after complaints of locals. After investigating sample water, EPA found that ground water contained compounds likely associated with natural gas production. The draft report was published in December 2011 (US Environmental Protection Agency 2011a). The second case stems from Dimock, Pennsylvania, where in November 2011 EPA announced that in four investigated home wells inorganic hazardous substances were found that present a public health concern. The following memorandum, which was published in January 2012, reported the presence of barium, DEHP, and glycol compounds, manganese, arsenic, phenol and sodium, all known to be used in hydraulic fracturing processes (US Environmental Protection Agency 2012b).

On the request of US Congress the EPA has also been working on a broad study on the impacts of hydraulic fracturing on drinking water quality, of which final results had not yet been published at the time of writing (US

Environmental Protection Agency 2011b). Meanwhile, what has widely become known as the "Halliburton loophole" continues to ensure that comprehensive federal regulation of hydraulic fracturing as it relates to potential groundwater contamination remains elusive. In short, an insertion in the Energy Policy Act of 2005 amended the Safe Drinking Water Act to exempt hydraulic fracturing as a technology from the so-called Underground Injection Control (UIC) program, except when diesel fuel was used in the process (US Government Accountability Office 2012).[17] Generally, EPA uses this UIC program to regulate the injection of fluids underground, but the exemption made it impossible for EPA to regulate potential groundwater contamination caused by fracking.

Next to fluids used in hydraulic fracturing, natural gas itself can contaminate ground water. A study on the Marcellus and Utica shale rock formations in Pennsylvania concluded that there was systematic evidence for methane contamination of shallow drinking water systems in at least three areas where hydraulic fracturing occurred as well. In 85 percent of the wells under study methane concentrations were reported, but they were substantially higher when closer to natural gas wells. The same study found no evidence for contamination of ground water with fluids used in hydraulic fracturing (Osborn *et al.* 2011). In January 2013 the US Geological Survey published a study examining the water quality of shallow domestic water wells in northern Arkansas, focusing on chloride and methane concentrations, in which no evidence was found of groundwater contamination linked to the gas industry (Kresse *et al.* 2012).

Vengosh *et al.* (2014) found that data published in several studies until January 2014 contained evidence of stray gas contamination (which is the contamination of shallow aquifers with fugitive hydrocarbon gases), surface water impacts in areas of intense shale gas development (e.g. because of leaks or spills), and the accumulation of radium isotopes in some disposal and spill sites. However, the direct contamination of groundwater from hydraulic fracturing fluids and deep formation waters by hydraulic fracturing itself remain disputed. The authors also note that these risks can be mitigated with increased engineering controls during well construction and alternative water management and water disposal options (*ibid.*: 8847). In sum, with no clear empirical outcome at this stage it has been difficult to make general statements about relations between shale gas extraction and drinking water quality.

Induced seismic activity

During 2011, the Youngstown, Ohio area experienced twelve seismic events ranging from 2.1 to 4.0 magnitude on Richter scale, according to a study by the Ohio Department of Natural Resources (2012). Each of these events occurred into a mile radius of a so-called class II deep injection well, being the type of well that oil and gas producers use to dispose injection fluids.

While disposal wells are not the same as hydraulically fractured gas wells, the increase in deep water injection wells during the last years in Ohio is directly linked to the industry's need to dispose flow-back fluids from shale gas operations in Ohio and neighboring Pennsylvania, where this practice has been prohibited since May 2011 (*ibid.*). The conclusion of the report is that it is probable that the earth tremors were induced (in this context meaning that the tremors are the result of human activity).

From January 2011 onward the Oklahoma Geological Survey registered 50 small tremors from 1.0 to 2.8 magnitude nearby shale gas extraction operations, but was unable to say "with a high degree of certainty" that these were induced (Holland 2011). Another study linked seismic activity in north central Arkansas to waste fluid injection from hydraulic fracturing operations (Horton 2012). The US Geological Survey published a study that documents a seven fold increase in seismic activity in central US since 2008, largely associating this increase in seismic activity to the large increase in the number of waste water disposal well injections (Ellsworth *et al.* 2012). In June 2012, however, the National Research Council pre published its results of an examination of scale, scope and consequences of induced seismicity during fluid injection and withdrawal activities related to among others shale gas extraction. The authors concluded that the process of hydraulic fracturing does not pose a high risk for induced seismicity (National Research Council 2012: 156). In addition injection of waste water derived from energy technologies such as hydraulic fracturing does pose some risk for induced seismicity, but "very few events have been documented over the past several decades relative to the large number of disposal wells in operation" (*ibid.*). However, Keranen *et al.* (2013) studied the impacts of wastewater injection and their links to induced seismic activity in Oklahoma. Their results suggested that time delays can exist between the start of injection and induced earthquakes. In the case of Oklahoma the first noted earthquake (M_w 4.1, 2010) took place 17 years after injection started. Over time, with sufficient volume injected and wellhead pressure, the pressure at the fault may exceed critical pressure and trigger a seismic event. The time it takes for pressure at the fault to pass a critical pressure point depends on the injection rate and permeability of the rock. Though some caveats with regard to the data remain, the authors concluded that one of the M_w 5.0 earthquakes in Oklahoma had been induced by increased fluid pressure.

In December 2012, in the United Kingdom the shale gas industry received the green light to resume operations after research had been concluded into two seismic events near Blackpool in 2011 with 2.3 and 1.5 magnitude. In June 2012 the Royal Society and the Royal Academy of Engineering published their review of scientific and engineering evidence regarding risks associated with hydraulic fracturing. It concluded that "seismic risks are low" and leaves the carbon footprint of shale gas as the only contentious issue related to shale gas extraction on the table for further research (Royal Society and Royal Academy of Engineering 2012). Though geologists determined it was highly probable that the seismic events in Blackpool were induced operations were allowed to be resumed with the provision that a traffic light system be implemented to monitor operations. Under the new British regulations, as of December 2012, operators seeking to explore shale gas through hydraulic fracturing have to do substantial research prior to operations regarding seismic risks, submit a plan how these risks are addressed and carry out seismic monitoring before, during and after operations. Also, a "traffic light system" is required to monitor unusual seismic activity.[18] Earlier reports suggested that this system would require companies to halt operations when new seismic activity would exceed 0.5 magnitude, but more research is needed to study the new British regulations and their impact on shale gas operations and industry's activity.[19]

Arguably a number of industrial activities has been linked to induced seismicity, including reservoir impoundment, mining, construction, waste disposal, and perhaps most prominently in recent years, fluid injections for geothermal energy exploitation (Majer *et al.* 2007). Up to date the US Department of Energy states that most of the induced seismic activity qualifies as an "annoyance," not a risk, and argues that proper engineering can minimize the chances of seismic activity. Yet so far it is unclear where the threshold between acceptable nuisance and unacceptable risk is. In addition it is unclear whether and if so what from a regulatory perspective can be done to minimize the risk, for an earthquake of magnitude around 4.0 and up (that have repeatedly been reported in central US) seems like a nuisance one would want to avoid in more densely populated areas. Perchance the newly launched British regulatory model could be used elsewhere, though with controversy surrounding the topic in the US it seems unlikely that happens any time soon.

Water availability and recycling

Water issues linked to shale gas extraction also involve the availability of the resource. The EPA estimates that the water quantity needed to fracture a horizontal well can go up to 5 million gallons, depending on depth, horizontal distance and the number of repeated operations (US Environmental Protection Agency 2010). In areas where water is abundant, such as Pennsylvania, this is not an issue. However, in drought-prone regions, such as Texas, water availability can be an issue of concern (see also Vengosh *et al.* 2014). In July 2011, the Texas Water Development Board estimated that the shale gas industry used about 12 billion gallons of water per year in Texas, a number that was expected to grow up to 40 billion gallons per year in 2030 (Nicot *et al.* 2011). A study focusing on the three major shale gas plays in Texas to quantify the net water usage for shale gas production, found that roughly 10 percent of annual water use in Dallas is destined for the shale gas industry. For the whole state of Texas, water usage for shale gas is under 1 percent of total withdrawal, however, local impacts vary with the availability of water and competing demands (Nicot and Scanlon 2012). Freyman and Salmon (2013) observed that in Wise and Johnson counties in Texas, in 2011 water demands for hydraulic fracturing operations represented respectively 19 percent and 29 percent of the overall county water use. Some authors suggested that the debate on water availability in places such as Dallas/Fort Worth, Texas, despite the significant concern among local citizens, diverts attentions from the primary factor affecting water supply in expanding urban areas: increasing municipal water use (Fry *et al.* 2012). As fresh water becomes scarcer and prices of water rise, it is also worth noting that the industry has started developing alternatives, such as the use of brackish or saline water, which is of no interest for civilian or agricultural use because there is too much salt in it. Freyman and Salmon (2013) concluded that in the Eagle Ford basin in Texas up to 20 percent of the water used for hydraulic fracturing is saline water.

Another environmental concern linked to shale gas and water is what to do with wastewater, once it has been injected into the well together with sand and chemicals. The most quoted options are reinjection in the well (that has been linked to induced seismic activity), discharge to surface water after treatment or application to land surfaces. Data from 2011 from the Pennsylvania Department of Environmental Protection suggest that about half the wastewater was treated; about one third was recycled to be used in other hydraulic fracturing operations, while less than one tenth was injected into disposal wells (Hammer and VanBriesen 2012). Wastewater handling has been reported as a key problem for environmental opposition in several cases (Rahm 2011). Some studies have suggested that federal and state regulations have not kept pace with the shale gas industry and should be strengthened to reduce risks of hydraulic fracturing for current regulatory frameworks are "inadequate" to do so (Hammer and VanBriesen 2012). Shariq (2013) noted that even the best waste water treatment technologies cannot strip all toxic chemicals from the water and are often selectively implemented because of costs. Alternative fracturing fluids and the use of non-fresh water are part of ongoing research activities (Pearson *et al.* 2012). In Canada, over 1.200 successful simulations have taken place of what is called Dry Frac, a process that uses liquid CO_2 as the carrier fluid in fracturing operations without using water or any additional treatment additives. One challenge to overcome is the formation of ice in drilled wells, which can be done using N_2 gas (nitrogen). One key challenge remaining for these inexpensive fracturing fluids to become serious commercial alternatives is the lack of infrastructure to transport N_2 and CO_2 (Kargbo *et al.* 2010).

Regulation

Following these environmental concerns an obvious question is if, and if so, how these issues can be addressed to avoid environmental disruptions or worse. Here, the US and the EU seem to differ fundamentally. That difference appears to originate from what is called the precautionary principle, as laid down in the EU Lisbon Treaty (European Union 2010a). In short, the principle aims to ensure a higher level of environmental protection through preventive decision-taking in case of risk. It is used, in particular, where scientific data do not permit a complete evaluation of the risk and can then be used to stop or withdraw products or services considered to be potentially hazardous. The absence of this principle in US law arguably contributed to the situation in which industry has taken the lead regarding hydraulic fracturing and shale gas extraction, while legislature and regulatory authorities on both federal and state levels have been trying to keep pace, as has been demonstrated by several examples of environmental concerns in the previous sections.

United States

Overall the primary regulatory authority for shale gas is at the state level. The lack of federal regulations in most issues related to shale gas extraction has resulted in a wide variety of approaches towards current practices in the country, varying from warm embracement of technology and further exploitation of natural gas (e.g. Texas) to outright bans (Vermont, New York). Efforts to regulate some of the environmental concerns have mostly occurred on the state level and have in few instances even encountered outright hostility, such as in Texas (Rahm 2011). It is worth noting, however, that in Texas too regulatory evolution can be observed, as for instance demonstrated by the more stringent rules regarding the disclosure of chemicals used in hydraulic fracturing, as adopted in 2012.[20]

The disclosure of chemical constituents used in hydraulic fracturing fluids is still not always required under federal and most state laws, even though significant developments are identified.[21] While industry representatives first claimed this information to be proprietary, an increasing number of states have installed regulations that force gas companies to disclose either what chemicals are used or what quantities of chemicals are used. Yet most of these regulations contain trade secret exemptions. The first exception is proposed regulation in Alaska, where the Oil and Gas Conservation Committee proposed rules without these exemptions. It remains to be seen whether these proposals make it into law.[22] Again, significant differences between states have been reported, in some cases linked to states being "energy dominant" such as Texas or not, in this analysis Colorado (Davis 2012). Attempts to regulate disclosure of chemicals on the federal level, for example, in the form of the Fracturing Responsibility and Awareness of Chemicals Act (also referred to as the FRAC Act) that was introduced to the US House and Senate in June 2009, have so far failed.[23] In May 2012 the US Department of the Interior published proposed rules for gas companies working on public and Indian lands that require the disclosure of chemicals used in hydraulic fracturing operations, yet only after operations have been completed (US Department of the Interior 2012).

Water quality protection on the federal level is arranged under the Safe Drinking Water Act, which prohibits the underground injection of fluids from endangering drinking water. However, hydraulic fracturing operations have been excluded from these regulations under a 2005 provision amending the Safe Drinking Water Act, the only exception being hydraulic fracturing operations involving the usage of diesel.[24] The federal EPA is investigating the impacts of hydraulic fracturing on drinking water, by examining the effects of large volume extraction of water, chemical mixing, well injection, flow-back and produced water, and waste water treatment and water disposal (US Environmental Protection Agency 2012d). While the progress report in late 2012 did not contain conclusions, final results have not been published at the time of writing. Existing attempts to regulate water quality on the federal level comprises the Fracturing Responsibility and Awareness of Chemicals Act, yet passage appears unlikely in the nearby future, according to some because the EPA study results shall be awaited (Jackson *et al.* 2011).

The treatment, disposal and reuse of wastewater is subject to several regulations, though not adequately protective according to some (Hammer and VanBriesen 2012). Note that discharge of wastewater into surface water without treatment is not allowed. However, as a consequence of its exception from the Safe Drinking Water Act, if wastewater is treated for the sole purpose of reuse in hydraulic fracturing operations, it is not subject to federal regulation. On the state level there is authority to regulate these issues, as occurs in some cases.

One area in which federal regulation has been adopted is air quality. In April 2012 the federal EPA used its authority under the Clean Air Act to regulate emissions from drilling activity. From 2015 onward gas producers have to abide to federal rules for natural gas wells that are hydraulically fractured, demanding these companies to apply what have been called Reduced Emissions Completions (e.g. the application of capture technology to prevent damaging gases, such as volatile organic compounds or methane, from coming into the air). Until 2015 companies are required to flare these emissions, while venting is prohibited (US Environmental Protection Agency 2012c). Further research is needed to identify what share of the wells in the US in fact fall under these regulations, for several exemptions apply. It is also worth noting that seven states (New York, Maryland, Delaware, Vermont, Connecticut, Rhode Island, and Massachusetts) announced in late December 2012 that they would sue the federal EPA, seeking to force the agency to regulate methane emissions from natural gas operations. According to the states the mentioned air quality regulations are not sufficient to address methane emissions from shale gas operations and stricter regulations are needed here.[25]

European Union

Despite the fact that not a single molecule of shale gas has been produced in EU, supranational regulatory development is significant, notably partly based on US experiences with regard to environmental issues (European Parliament 2011). In early 2012 the EC published a commissioned report on the existing legislative framework in the EU, which examined four of its member states (i.e. Germany, Sweden, France, and Poland; Philippe & Partners 2011). The authors concluded that the regulatory framework was sufficient for the early exploratory phase shale gas extraction in the EU is in, but it can be debated whether that is the relevant question. The report does not assess whether all existing legislation has been transposed into national law. Furthermore it seemed relevant to know whether moving beyond the exploratory phase of extraction is possible within the existing regulatory framework.

In September 2012 the EC published three sizeable reports about the impacts of shale gas on markets, environment and climate (AEA 2012a, 2012b; Pearson *et al.* 2012). The report on the environmental impacts of shale gas extraction concluded that shale gas extraction generally imposes a larger environmental footprint than conventional gas extraction (AEA 2012a). It identified environmental pressures in terms of land-take, releases to air, noise pollution, surface and groundwater contamination, water resources, biodiversity impacts, traffic, visual impact and seismicity. Currently 19 pieces of EU legislation are relevant to all or some of the stages of shale gas extraction.[26] The report also listed a number of gaps in existing legislation and some potential gaps, most notably that Environmental Impact Assessments are not required and that there may be potential gaps regarding waste management, emissions to air, water contamination, water use and noise.

The study on climate impacts of shale gas extraction suggested that greenhouse gas emissions linked to electricity produced by burning shale gas are 2 percent to 10 percent lower than emissions from electricity generated from sources of conventional pipeline gas located outside of Europe, and 7 percent to 10 percent lower than that of electricity generated from LNG (AEA 2012b). However, these results evaporate when emissions from shale gas during the entire extraction phase are not effectively controlled ("venting"). Questions remain as to what existing framework would be appropriate to regulate emissions from shale gas extraction. The authors name the Environmental Impact Directive, the Directive on Industrial Emissions and the European Emissions Trading Scheme as possible vehicles. McGowan (2014) noted that given the momentum of shale gas in the EU and the extensive environmental mandate, it might be expected that a coherent response to shale gas would be both "necessary and desirable."

Even though in late 2013 it was generally assumed that the EC would propose additional binding legislation regarding shale gas extraction in the EU, this eventually did not happen. Boersma and Khodabakhsh (2014) described how several member states, in particular the United Kingdom and Poland, openly pressed EC President Barroso to tone down EC ambitions. Thus, in January 2014 the EC published a non-binding Recommendation on minimum principles for the exploration and production of hydrocarbon using hydraulic fracturing.[27] The recommendation among others urged member states to carry out strategic environmental impact assessments prior to granting licenses for hydraulic fracturing operations, to draft baseline reports about for instance air and water quality, proper well closure, and limited venting and flaring and maximum methane capture. In essence, the EC will monitor the implementation of its recommendation in the first 18 months after publication, keep public scorecards, and if necessary come up with additional legislation after the monitoring period.

It is worth mentioning that even regulations adopted by European institutions provide no safeguard as such. History has numerous examples of directives not being implemented timely and/or properly by EU member states. To give an example, under current circumstances the Polish authorities have not fully implemented all existing European guidelines related to shale gas (e.g. the Waste Directive or the Directive on inland transport of dangerous goods). Furthermore, several issues are not part of the regulatory framework but, referring to ongoing debates in the US, would deserve attention in case of commercial exploitation of shale gas. The most obvious examples of this are the lack of specific requirements regarding the prevention of contamination of ground and surface water or the absence of specific disclosure procedures on hydraulic fracturing fluids. Next to environmental and procedural regulations, Poland has to meet the obligations as laid down in the European legislation for the internal market of gas (European Union 2009). Though in theory the Polish market is open to competition since 2007, some serious obstacles remain. State-owned incumbent company PGNiG represents over 90 percent of gas sales in the country and is also responsible for all distribution networks in the country. Therefore a competitive gas market in the years ahead is difficult to envisage. Also, it remains to be seen when natural gas price regulation for both industry and end consumers will be abandoned. These problems were reiterated in 2012, when the EC published an overview of pending infringement procedures, one of them against Poland for not properly implementing both the Second and the Third legislative package (European Commission 2012).

So what if regulations have not been implemented? Formally, under the Treaty of the Functioning of the European Union, the EC guards over proper implementation of European law in the member states and can start infringement proceedings in case of non-compliance. Ultimately it may also refer the case to the European Court of Justice. History has shown that the Commission sometimes struggles to compel member states to implement European law, despite its formal powers to do so. This issue is further addressed in Chapter 7 of this book.

Gas infrastructure

Infrastructure is crucial to the functioning of any gas market and it has been argued that the organization of infrastructure in the US contributed to a large extent to the developments regarding shale gas (see e.g. Medlock 2012b). This section will touch upon these issues and the remaining infrastructural challenges in the US, before turning to the EU. Here the focus shall be on Poland, where in the years ahead shale gas extraction – if at all to happen in Europe – is expected to occur. The argument made here, bluntly that Poland is not ready for shale gas extraction, does not necessarily add up for other European member states that have shale gas resources under their soils. Yet ironically, most member states that have well developed infrastructure do not appear to have an interest to proceed with shale gas extraction on short notice.

United States

Market structure is an important characteristic to consider when assessing why shale gas production in the US has taken off the way it has. Some labeled it "most underappreciated factor that positively benefited shale gas production" (Medlock 2012b). Arguably small producers played a significant role in the push for shale gas production that the US has undergone during the last decade. One characteristic typical to the US market that has contributed to this development, is the unbundling from capacity rights from pipeline ownership (*ibid.*). This means that the owner of a pipeline cannot also own natural gas flowing through it. Without a "personal" interest of the pipeline operator, any company can access the market through competitive bids. This is different to market structures on the other side of the Atlantic Ocean, where small producers may in fact be hindered to access the markets.

The only issue related to infrastructure in the US that has been fiercely debated relates to expanding gas infrastructure investments to facilitate the increasing share of natural gas for expanded gas-fired electricity generation.[28] With abundant and cheap natural gas being a favorable fuel to generate electricity and the foresight of cheap natural gas being an incentive to invest in additional gas-fired electricity generation capacity, the open question is who should pay for the infrastructure linking these pipelines to the grid. As evidence that it is difficult if not impossible to keep pace with the US market, it is worth noting that infrastructural bottlenecks continue to cause problems in parts of the US. In early 2014, the lack of physical infrastructure to transport more natural gas for electricity generation into New England contributed to significant price spikes for natural gas, and ironically even an occasional shipment of LNG was brought into the US to facilitate peak demand.[29]

European Union

Poland is a gas country under construction. The share of natural gas is limited, with a yearly domestic consumption of around 14 bcm, forming roughly 13 percent of the country's primary energy consumption. Of this gas, two thirds are imported, exclusively from Russia. Given the limited role of natural gas, it cannot be a surprise that gas infrastructure is not substantially developed. To give an example, only 54.6 percent of Polish households is currently connected to gas networks (Central Statistical Office (Poland) 2012). Most pipelines are located in the southwest of the country, where industry is clustered, and around the main urban areas, but not necessarily in the areas where shale gas would be produced. Furthermore large transit pipelines have been built across the country from east to west.

Following the potential extraction of shale gas resources and continued high dependence on Russian gas imports, increased investments in Polish gas infrastructure are made. The demand forecasts used by the national transmission system operator, called Gaz-System, show that maximum gas demand in Poland is expected to double within this decade, to 30 bcm in 2021. First, Gaz-System has been investing in interconnection capacity to get the Polish gas market out of its isolation and connect it to neighboring countries to its west and south. The existing interconnector with Germany in Lasów has been upgraded to a maximum capacity of 1.5 bcm starting January 2012. To the south an interconnector has been launched in September 2011 on the border with Czech Republic at Cieszyn with a capacity of 0.5 bcm, albeit not yet two-directional. These investments are part of the Transmission System Development Program for 2010–2014 in which 1000 kilometers of new pipeline are envisaged.

Next to these commissioned projects, several interconnections are under study. In January 2012, Gaz-System and its Lithuanian counterpart Lietuvos started a feasibility study on an interconnector between the two countries. In August 2014 ACER published its decision on the cost allocation for this project, as Poland and the Baltic states could not reach an agreement, as discussed earlier.[30] Gaz-System and Eustream are currently preparing an interconnector between Poland and Slovakia.[31] In 2013 both existing interconnectors with Germany and Czech Republic received green lights for further upgrade as well. Both these projects were substantially financed through the European Energy Program for Recovery (European Union 2010b), representing €10.5 million for the Czech interconnector and €14.5 million for the German interconnector.

Several other projects could contribute to further diversification of gas supplies in Poland. One example is an LNG terminal that is constructed in Świnoujście in the northwest of the country. Its maximum capacity is 5 bcm at a cost of €700 million, of which roughly half is financed by the EC.[32] In addition the European Investment Bank has loaned Poland approximately

€135 million to realize this project (European Investment Bank 2011). The terminal is expected to come on stream in 2015, though it is not without controversy. The Polish have contracted LNG from Qatar for their terminal, and although details of the agreement have not been released, several reports have suggested that the contracted LNG would be at least 30 percent more expensive than Russian natural gas.[33] Anecdotal evidence suggests that indeed a higher price is paid for LNG than current contracts, and the Polish authorities are rumored to plan to regulate the prices of LNG that is imported from 2015 onwards, because otherwise nobody would be interested to buy the natural gas. If confirmed, this would be the first case of a European country willing to pay a premium to diminish the share of Russian natural gas, and indirectly the Polish tax payer would pay a premium for a concept called energy security. The long term consequences of decisions like these deserve further analysis. Next to the LNG terminal the so-called Baltic Pipe is in its pre-construction phase, eventually aiming to link Poland and Denmark. It is intended to give Poland access to Norwegian gas, while the Danish have expressed interest in receiving Russian gas through Poland. The EC wants to invest €150 million in this pipeline (European Commission 2009).

And then there may be shale gas. The verdict of private companies is out, even though several companies have by now left the country, due to disappointing geology and disappointing governance. In addition, there are likely infrastructural hurdles to be overcome. Considering the geographic location of shale gas resources, spread out from the north around Gdansk to the Ukrainian border in the southeast, additional investments in infrastructure may be necessary to ship large amounts of shale gas, either domestically or internationally. Even without those additional investments, ongoing infrastructural projects may well occupy at least the first part of this decade. With substantial new supplies coming online from then onward – with for instance the commissioning of the LNG terminal in Świnoujście – the pressure is on for Polish transmission system operators and policy makers to prepare its domestic market for either gas consumption or significant gas imports and exports.

Energy security

While industry balances geologic and economic realities to assess local potential for shale gas extraction, political and academic debates have suggested another pressing agenda: energy security. In the subsequent sections, this is eluded to in both the US case and the European one. Again, the European case will focus on the example of Poland, where security considerations are high on the agenda and play an important if not decisive role.

United States

Roughly since 2000 it became clear that US natural gas consumption could no longer be stilled with domestically produced gas. In order to attract foreign gas, major companies invested in LNG terminals. In the meantime, however, smaller independent enterprises sparked the domestic quest for unconventional gas, among others shale gas. A decade later, the boom turns out to be enormous and it seems increasingly safe to state that by 2016 the US could be a net exporter of natural gas (US Energy Information Administration 2012a, 2014).

Not everybody likes that idea though. In early 2012 the Committee on Natural Resources of the US House of Representatives urged Energy Secretary Chu to investigate the consequences of exporting liquefied natural gas (LNG), because of worries that exporting natural gas would raise domestic energy costs, reduce business's global competitiveness, make the country more reliant on foreign sources of energy and slow the transition to a low carbon economy.[34] Although it is likely that the US in due time will become a substantial exporter of LNG (as touched upon earlier in this chapter), the debate on future LNG exports is ongoing at the time of writing. Below (2013) observes that over the last decade both US Congress and different Presidents have been applying comparable definitions of energy security. These typically include relying on diverse sources of supply, not being overly reliant on foreign sources, as well as ensuring a reliable energy infrastructure. Within discussions about energy independence, however, which blossomed with the surge in domestic energy production, both the legislative and the executive branch in the US place more weight on increasing production over reducing consumption (*ibid.*: 866).

European Union

In Poland energy security is rooted in profound distrust of Russia and even Poland's fellow EU member states on questions of energy resources. This is perhaps not surprising, given the history of war, subjugation, hegemony, and mistrust in this part of Europe, historically often under the yoke of Russia and Germany. But alongside the tragedies of history is a seemingly unacknowledged reality that Russian companies have been stable suppliers of both natural gas and crude oil to Poland for many decades. During the oft-cited price dispute between Russia and Ukraine in January 2009, which caused supply interruptions to European consumers and caused some in southeastern Europe to actually go without gas for several days, Gazprom actually increased shipments substantially to Europe via the Yamal pipeline (which traverses Belarus and Poland), so that consumers in Poland and Germany did not feel the interruption (Le Coq and Paltseva 2012).

Dependency on Russia has been a heated topic of discussion in the EU, and Poland in particular. Some have even speculated that the development of unconventional gas resources in Europe is enabled by the unpredictability of Russian supplies (Kuhn and Umbach 2011). The inauguration of the Nord Stream pipeline linking Russia and Germany beneath the Baltic Sea represents a strategy by governments, Gazprom, and western European industry to reduce transit risk by bypassing intermediary countries such as Ukraine and Belarus. The reaction in Poland to the Nord Stream pipeline was negative, since it was also bypassed by the pipeline. It urged some observers to recall the past, calling Nord Stream the Molotov-Ribbentrop pipeline (as described by Le Coq and Paltseva 2012) and even the then defense minister Radek Sikorski made allusion to the Nazi-Soviet non-aggression pact in 2006 (Roth 2011). Since the inauguration of the Nord Stream pipeline, there has been a great fear in Poland that Russia would be in a better position to use gas as a political instrument.

Yet there are some things worth noting. Several assessments of the risk to EU gas supply have highlighted the lack of interconnectivity within the EU, and not Russian aggression, as being the main transit risk (Noël 2009; Le Coq and Paltseva 2012). Contrary to a commonly held view, transit risk did not increase in the period 1998–2008 (Le Coq and Paltseva 2012). While the Kremlin may have used its energy export capabilities as a short-term leveraging tool in the past, historical examples of this underscore the point made by Larsson (2006) that supply interruptions targeted at Poland or another EU member state are highly unlikely, as are long-term cutoffs that would impact Poland or another EU member state for an extended period of time.

According to some, energy policy discourse in Poland exhibits a high degree of "securitization," such that the topic of energy is often framed in terms of national security and an existential threat (Roth 2011). As discussed in Chapter 2, others have questioned whether energy security, certainly in other parts of Europe, has not merely been politicized (McGowan 2011; see also Kratochvíl and Tichý 2013). It is also unclear why, if Russian natural gas is perceived as an existential threat, Polish authorities have not done more to develop alternatives to counter this potential threat. Investments could have been made in for instance a pipeline with Denmark, connecting Poland to large reserves of alternatives of Norwegian gas, or in interconnection and reverse flow facilities with Germany, so that the country would have been connected to alternative supplies from northwestern Europe. The fact that Poland did not do this seems to be evidence that this is not an example of securitization in the way the Copenhagen School perceives it. More research is needed to assess why Polish policy makers have not developed alternatives for Russian gas. Nevertheless, the security aspects of the current shale gas discussions in Poland are noteworthy, and the debate stands in contrast to energy discussions in other parts of Europe. Values and beliefs, so important in shaping social orders according to North (1991) also vary substantially in Europe, this example demonstrates. Since joining the EU in 2004, Poland has been a strong voice in bringing energy to the fore of discussions of European external relations amid widespread perceptions in central and eastern Europe that energy policy in the EU prior to the 2004 enlargement had not sufficiently addressed Europe's overdependence on energy imports (Roth 2011). It was largely Polish efforts that led to "energy solidarity" language being inserted into the Lisbon Treaty. As discussed, Poland has actively – and arguably quite successfully – lobbied to get European funds to build its energy infrastructure.

While the framing of shale gas in security terms in certain circles of the Polish elite is perhaps understandable, it is also very likely counterproductive given the actual administrative and legislative hurdles that could impede the development of this resource. Moreover, in lack of coordination, Polish efforts to excite other member states for future shale gas extraction have appeared rather opportunistic. Just before the country became the President of the EU it suggested that shale gas extraction should be a project of common interest, while six months later this position had shifted (Wyciszkiewicz *et al.* 2011). It is worth noting that over the summer France had decided to ban hydraulic fracturing, while other member states had announced further research into environmental concerns. To stress the dubious nature of this situation, the EC reiterated that it "remains neutral" as regards member states decisions concerning their energy mix.[35] It is unclear why the Polish chose to have this confusing discussion in the first place, for European institutions have historically not been involved with choices relating to member states' energy mix, as reiterated in the Lisbon Treaty.

Polish officials often argue that Poland's dependence on Russian gas supplies is risky. Even if Russia was not a dependable supplier – which is questionable as just shown – it remains unclear why that would be a problem for a country that uses almost exclusively oil and coal as primary energy resources. Hence the irony is that current attempts of Polish transmission system operator Gaz-System, Polish government officials, European institutions and all others involved in the development of the Polish gas market, are in fact likely to increase future dependence on its most feared neighbor, unless the aforementioned trend of paying a premium for natural gas in the form of LNG is expanded in the future. Even when assumed that Polish shale gas will be extracted at some point, the most recent geologic forecasts demonstrate that there will only be sufficient supplies to cover a few decades. It makes sense to expect that at some point not just shale gas will run through those pipelines, but also increased LNG supplies, or Russian gas. In that sense the Polish lobby to extract shale gas and develop its gas infrastructure hence may well increase its dependence on Russia, instead of decreasing it.

It is worth considering developments in the EU in 2014 when discussing energy security considerations. Kratochvíl and Tichý (2013) noted that historically the prominent energy discourse in the EU has been the integration discourse, emphasizing the mutual benefits derived from energy cooperation between the EU and Russia. In light of the aforementioned nuances regarding energy security in central and eastern Europe, it is questionable whether in fact one EU discourse on energy was ever in place. Arguably the cooperation and integration discourse has featured prominently in the larger part of Europe (i.e. the EU 15), but less so in eight of the ten member states that joined the EU in 2004 (all except Malta and Cyprus), as the case of Poland demonstrates. In light of the ongoing crisis in Ukraine several questions cannot be answered at the time of writing, but are worth taking into consideration. First, as the case of Lithuanian LNG seems to provide a second example of a member state that is willing to pay a premium for natural gas that is non-Russian, what do these individual efforts mean in terms of European integration and collaboration? Perchance they should be seen as an expensive insurance policy against market power abuse? Furthermore, in light of global gas market developments, including those in North America, it seems likely that market dynamics have fundamentally altered the fundamentals underlying the traditional Russian–European energy relation (see also Kropatcheva 2014). Empirical evidence suggests that absent very drastic policy interventions, Russia is likely going to be a prominent supplier of natural gas to the EU for decades to come (e.g. Boersma *et al.* 2014). Having said that, the Ukraine crisis did substantially raise the profile of energy relations with Russia, and the new EC under President Juncker is expected to present its plans for an Energy Union in 2015.

Multilevel governance framework with institutional overview

Tables 6.1 and 6.2 contain an institutional overview of decision-making structures in both the US and the EU regarding shale gas. It considers production, distribution and regulation. The findings are discussed below.

Discussion

Geologic realities are a crucial factor in the EU with regard to eventual shale gas extraction, similar to North America or any other part of the world. Yet environmental concerns that have been linked to shale gas extraction and hydraulic fracturing could, in contrast to the US, potentially halt extraction in the EU before it has started. Within existing EU governance structures, policies on energy supply and production are predominantly the purview of the individual member states, though increasingly European institutions become more important stakeholders. Policies on environmental, water and air quality on the other hand are mostly developed in Brussels. The brief shale gas history in the US suggests that environmental regulatory authorities, policy makers and academia were caught off-guard and have been catching up with industry power-play since. Even today in the US it appears to be extremely difficult to institutionalize environmental regulation at the

Table 6.1 United States decision making regarding shale gas

Governance level	Public domain	Private domain
Federal	US Environmental Protection Agency • regulates air quality from 2015 onward • investigates impacts of hydraulic fracturing on water, unclear whether eventually decisions shall be taken at federal or state level. Department of Interior installed disclosure measures for hydraulic fracturing operations on federal and Indian lands (roughly 20% of US gas production in 2012). Federal lands are administered by, for example, National Park Service, US Forest Service and Bureau of Indian Affairs. FERC decides over interstate gas pipelines (see Chapter 5).	Investments in interstate gas pipelines are made by private consortia, usually operators and shippers.
Interstate	Water-related issues (conservation, utilization, withdrawal, development and control) are regionally decided upon in regional governmental agencies, where governors of the relevant states and representatives of the federal government take part as well. See for instance Susquehanna River Basin Committee (Pennsylvania, NY State and Maryland) and the Delaware River Basin Committee (Pennsylvania, NY State, New Jersey and Delaware).	

Table 6.1 continued

Governance level	Public domain	Private domain
State[a]	The primary regulatory authority for shale gas extraction is on the state level. Only air quality and disclosure rules on federal lands are regulated from Washington, DC. State regulators (EPA, Texas Railroad Commission and many others) have initiated varying disclosure rules in some states. Other forms of regulation are designed here as well, for instance waste water treatment, intrastate pipelines, site abandonment, taxation, permitting, safety, etc. In most states a number of agencies have responsibility regarding the regulation of shale gas extraction (e.g. agencies for environmental protection, conservation, emergency management, transportation).	In particular private smaller companies are claimed to have played an important role in the early history of fracking. Later on larger companies (majors) came into the market.
Regional and local		Local citizens owning land have had a decisive say in shale gas extraction, and many did agree with the industry moving in, arguably motivated by financial motivations (selling land/property rights to gas companies). Roughly 80% of US gas production takes place on private lands. Some local communities have installed fracking bans.

Note: [a] US state regulations and the distribution of responsibilities varies from state to state. An accurate example of the broad spectrum of responsibilities and involved agencies is provided by Blohm *et al.* (2012: 362ff.). It is worth noting that their analysis does bypass mineral rights ownership, while in this analysis the local level is labeled as one of the important levels of decision making.

Table 6.2 European Union decision making regarding shale gas

Governance level	Public domain	Private domain
Supranational/EU	EC institutions are largely responsible for regulation of environmental pressures in terms of land-take, releases to air, noise pollution, surface and groundwater contamination, water resources, biodiversity impacts, traffic, visual impact and seismicity. EC does occasionally (co)finance gas infrastructure projects in the member states, though its financial means are limited.	
Interstate (EU member states)	Water related directives are sometimes coordinated on a regional scale, e.g. when large cross boundary rivers are concerned, yet implementation of law is national affair. Coordination may vary throughout the EU.	
State/EU member states	Implementation of EC directives is a national responsibility. Non-compliance can eventually result in arbitrage at the EU Court of Justice, but usually depends on context, number of member states being noncompliant, etc. Implementation of water directives is a national affair. Exploratory shale gas activities are done by both public (for instance PGNiG, Poland) and private companies. Planning and investment of infrastructure mostly occurs on national level, by both public and private companies. Incidentally cross border investment occurs as well (e.g. Netherlands–Germany, GTS). Contrary to the US ownership and mineral rights reside in the national public domain. Therefore, decision making on the local level is limited. Local resistance, however, can be substantial and important (e.g. United Kingdom). Taxation is also a national domain.	Exploratory shale gas activities are done by both public and private companies Planning and investment of infrastructure mostly occurs on national level, by both public and private companies. Incidentally cross border investment occurs as well (e.g. Netherlands–Germany, GTS).

Table 6.2 continued

Governance level	Public domain	Private domain
Regional and local	Limited formal decision-making powers, but influence in protest movements not to be ignored, see for instance United Kingdom, Netherlands, and Germany. Also, environmental permits may have to be issued at the regional and local level.	Though formally decision-making powers may be limited, influence of local players has been argued to play a role in the Dutch decision to postpone hydraulic fracturing operations. More examples of this are available throughout the EU.

federal level, as for instance suggested by the repeated suspension of the Fracturing Responsibility and Awareness of Chemicals Act or the watered-down version of the rules for chemicals disclosure on federal lands that has been published by the Department of the Interior. At the state level though substantial regulatory developments can be identified, though arguably this has been a complex and time-consuming process as well. To give an example, in the case of disclosure of chemicals used in hydraulic fracturing operations, several of the regulations that have been initiated do not seem optimal for their purpose, because they either require qualitative or quantitative data about the chemicals being used in operations, or contain several exemptions. In another example, Fry (2013) described how in many municipalities in the Dallas–Fort Worth area longer setback distances (i.e. physical distance between hydraulic fracturing operations and residents to protect their health, safety, and welfare) have been established than are required by Texas law (200 ft or 61 m.). It should also be noted that some of the most pressing environmental concerns are still under research. The lack of clear evidence on causes and consequences may make it increasingly difficult to regulate environmental risks, and perchance easier for lobby groups to push their interests. Yet while in the US the lack of clear evidence on some of the environmental concerns is evidently not a reason to halt operations, in Europe this may be different.

In pinpointing fundamental differences between the US and the EU regarding shale gas, several elements stand out. First, the market structure seems to have played an important role in the US with regard to shale gas development. As touched upon, the concept called "unbundling in the pipeline" and relatively easy market access for smaller companies as a result thereof, has enabled these smaller businesses to spur shale gas extraction on a large scale within a decade in the US. There is evidence that the larger companies had been investing in LNG terminals in the United States since the early 2000s, assuming that the US would rapidly become an importer of natural gas. Only at the end of the decade these companies realized that domestic production was exploding and subsequently they bought their way into the market (e.g. Exxon's purchase of Texas based XTO Energy for US$31 billion in 2009).[36] It is likely that the smaller gas companies have profited from low entrance barriers to the market, in combination with the lack of regulation of what was an unknown phenomenon at that time. More empirical research would be useful to determine exactly what the role of smaller and larger enterprises in the first decade of shale gas extraction in the US has been.

As appears from the institutional analysis, another important distinction between the EU and the US can be found on the local level. The US is unique in that resources found under the earth's soil are property of the land owner. Hence, with roughly 80 percent of current shale gas extraction taking place on private lands, land owners play a crucial role. Without their consent there is no shale gas extraction. That is a fundamental difference from any other country in the world, where the state usually owns whatever is in the ground. Of course, access to land can in the US be bought. Common features in lease contracts include signing bonuses, royalties, rents, and so on. Reports on the developments of these conditions show remarkable changes over time and differences per state. To give an example, in Pennsylvania in 2003, private landowners received about US$12 per acre in signing bonuses and a 12.5 percent royalty rate for shorter-term leases of five to seven years. In 2008, payments of nearly US$2.900 per acre and 17–18 percent royalty rates for these same leases were not uncommon (Andrews *et al.* 2009).

It would, however, be inaccurate to suggest that shale gas extraction has been embraced in full in the US. The different attitudes at the state level have been described before, and in fact show resemblance to the overall picture in the EU, the difference being that in the EU no member state has reached the commercial production stage. In addition, it is interesting to note that also in states where shale gas extraction takes place, some local communities have banned or continue to oppose hydraulic fracturing operations (e.g. the ongoing tug-of-war between institutions and local citizens in Fort Collins, Colorado).[37] In the November 2014 election round local communities in Ohio, California, and most notably Denton, Texas, approved bans on hydraulic fracturing.[38] And in December 2014 New York State made the controversial decision to ban fracking on its property.[39] In recent years, some research has been published about social acceptance of hydraulic fracturing. Boudet *et al.* (2014) found that to most respondents, energy extraction technologies are unknown territory, and there is therefore uncertainty whether to support it or not (see also Theodori *et al.* 2014). Kriesky *et al.* (2013) unsurprisingly found that support for hydraulic fracturing is larger where citizens reap direct economic benefits, for instance because they have a family-held land lease.

On the other hand of the spectrum, without this generous financial compensation, how would local citizens benefit from shale gas extraction? In France the lack of these local benefits has been mentioned – in combination with risks and nuisances linked to shale gas extraction – as one of the reasons to be against extracting shale gas resources and ban the technology of hydraulic fracturing (Leteurtrois *et al.* 2011: 44–45). But there has to be more to it than that. In an effort to spur shale gas extraction the British government in early 2014 proposed a benefit scheme and significant tax breaks for local communities in order to compensate those that would allow hydraulic fracturing operations.[40] Though effectively this could mean that millions of pounds would stay in local communities, this attempt to gain support received mixed responses. Arguably it may be too early to declare defeat, but since the announcement public opposition to hydraulic fracturing has not diminished, and anecdotal evidence suggests that government officials believe that the tax break may not be enough to gain public support.

As available evidence from Poland – arguably the European front line of shale gas extraction – suggests there can be several local barriers that hinder commercial extraction. First of all there are geologic realities, as confirmed by the substantial downgrades in recoverable reserves that the Polish Geological Institute published in spring 2012. Even if the geology is favorable, other factors are of importance. In the case of Poland the first is market development, in terms of building sufficient infrastructure, either for domestic consumption or for export, but also in terms of market access (e.g. shaping a market that is not monopolized by one state-owned company or end regulation of gas tariffs). Second, implementation of existing regulations and directives is required. Though – as Chapter 7 shows – there are many more examples of European member states being noncompliant with existing legislation and although the punitive route to the European Court of Justice is a time-consuming affair and its effectiveness in some cases can be questioned, at the end of the day the EC has a decisive say on most regulations that are related to environmental concerns that have been linked to shale gas extraction.

One element that has undoubtedly been of importance in shale gas discussions is the elusive concept of energy security. As shown in the analysis this concept seems to bring along many unfounded claims on both sides of the Atlantic Ocean. In Poland, it has been used repeatedly to spur the EC to express its support for shale gas extraction, even though historically the energy mix is an area left exclusively to the member states. As shown, these Polish efforts probably did not help the Polish case. In addition the case of the US suggests that a state approach in terms of getting large-scale shale gas extraction started can be rather effective and no federal intervention is required here.

Decision-making structures in this case study do not seem to differ greatly. On both sides of the Atlantic Ocean decisions regarding the extraction of energy resources are taken at the state level. As a result of this on both sides opponents and proponents of shale gas extraction are found, though the latter prevail in the US and are scarce in the EU. It is worth noting that on the federal level there appears to be more support for shale gas extraction in the US. One can think of the financial support for the development of technology in the 1970s (e.g. Boersma and Johnson 2013) or President Obama's speech when visiting Poland in 2011.[41] European institutions have been reluctant to express a position other than neutral regarding shale gas extraction, confirming the exclusive mandate of its member states, both those in favor (Poland and the United Kingdom) and those who oppose (most notably France and Bulgaria). In addition it is worth reiterating that although in the US the state decides whether shale gas extraction is allowed from under its soils, the local level too is of importance. This is due to the unique mineral rights legislation, which in a nutshell prescribes that the owner of land also owns what is underneath it. Hence landowners can lease their land to gas companies in order to extract natural gas and reap the – often substantial – financial benefits. In addition, local communities in several states have installed bans on hydraulic fracturing, though these are generally legally challenged at the state level.

In terms of environmental regulation decision-making structures in the US and EU are slightly different. European institutions have been more active in installing environmental regulations than their US counterparts have. A part of the explanation for this may lie in the fact that in the US "the market" is in the driving seat, instead of governmental and regulatory institutions. There may be several reasons for this. First, the currently preferred technology to extract shale gas has not been taken serious for a long time. Even though public private partnerships had been working to develop the technology since the early 1970s, it was not before the early 2000s when commercial extraction of shale gas took off. As discussed, it took the larger companies almost another decade to acknowledge the potential of this technology. Second, federal authorities in the US traditionally have been reserved to draft policies that can also be drafted on the state level. Resource extraction is considered to be such an affair. The exception is air quality, where the EPA has a mandate, but so far its attempts to regulate hydraulic fracturing and shale gas extraction have been hindered by legal exemptions and political obstruction. In Europe, the European Environmental Agency merely collects and analyses data, whereas European institutions draft regulations and member states are responsible for implementation. It is worth noting that sometimes in the US environmental regulatory authorities are viewed with skepticism and sometimes even outright hostility (see Rahm 2011). This is different from Europe, where environmental policy generally receives broad public support.[42]

Thus, from a European perspective this is a case *par excellence* for

European integration. Whereas resource extraction is an exclusive domain of the member states, regulation of environmental risks associated with resource extraction is predominantly a European affair. European institutions have repeatedly declared to be neutral with regard to shale gas extraction, a position seemingly aiming to satisfy all member states, from those who embrace shale gas to those that have legally banned it. In January 2014, however, the EC published a non-binding Recommendation with basic principles on shale gas extraction. Arguably, in 2013 the EC was considering more binding legislation for all member states, but several opposed heavily, as most prominently exemplified by the open letter of British Prime Minister Cameron to EC President Barroso.[43]

Acknowledgements

Earlier versions of this chapter have been written and published together with Professor Corey Johnson of the University of North Carolina, Greensboro, under auspices of the German Marshall Fund of the United States, based in Washington, DC. Special thanks therefore go out to Professor Johnson. In addition I would like to thank my colleagues of the Transatlantic Academy and those of the German Marshall Fund who have provided me with useful comments and feedback, and arranged the logistics during fieldwork in Pennsylvania, Brussels, and Poland.

Notes

1   See www.eia.gov/energy_in_brief/about_shale_gas.cfm.
2   A final decision to ban hydraulic fracturing in New York state was taken in December 2014. See www.reuters.com/article/2014/12/17/us-energy-fracking-newyork-idUSKBN0JV29Z20141217.
3   See www.examiner.com/article/just-six-pennsylvania-counties-account-for-the-majority-of-shale-gas-production.
4   See www.nytimes.com/2012/08/01/business/energy-environment/01iht-bp01.html?_r=0 (accessed October 10, 2012).
5   See http://mobile.bloomberg.com/news/2013-03-28/u-s-baker-hughes-gas-rig-count-declines-to-near-14-year-low-1-.html.
6   See www.bloomberg.com/news/2013-06-20/iea-says-u-s-natural-gas-output-to-accelerate-next-year.html.
7   See www.ferc.gov/media/news-releases/2014/2014-2/06-19-14-C-1.asp.
8   These countries are Australia, Bahrain, Canada, Chile, Dominican Republic, El Salvador, Guatemala, Honduras, Jordan, Mexico, Morocco, Nicaragua, Oman, Peru and Singapore.
9   See https://ustr.gov/trade-agreements/free-trade-agreements/korus-fta.
10 See http://turner.house.gov/news/documentsingle.aspx?DocumentID=319118.
11 It is estimated that by the fall of 2014 around 65 shale gas wells had been drilled in Poland. See www.euractiv.com/sections/poland-ambitious-achievers/shale-gas-poland-exploration-exploitation-308387.
12 See www.depreportingservices.state.pa.us/ReportServer/Pages/ReportViewer.aspx?/Oil_Gas/Wells_Drilled_By_County.
13 See www.naturalgaseurope.com/exxonmobil-leaves-poland-shale-gas.

14 See www.reuters.com/article/2013/05/08/poland-shale-idUSL6N0DP2WH2013 0508.
15 See www.reuters.com/article/2012/06/14/bulgaria-shale-idUSL5E8HEAL72012 0614.
16 See www.naturalgaseurope.com/czech-republic-plans-shale-gas-moratorium.
17 See also http://water.epa.gov/type/groundwater/uic/class2/hydraulicfracturing/wells_hydroreg.cfm.
18 For the official announcement, see www.decc.gov.uk/en/content/cms/meeting_energy/oil_gas/shale_gas/shale_gas.aspx.
19 See www.guardian.co.uk/environment/2012/apr/17/gas-fracking-gets-green-light.
20 See www.rrc.state.tx.us/about-us/resource-center/faqs/oil-gas-faqs/faq-hydraulic-fracturing.
21 For a review of state regulations, a study carried out by Resources for the Future is helpful: www.rff.org/centers/energy_and_climate_economics/Pages/Shale_Maps.aspx.
22 See www.eenews.net/public/energywire/2013/01/03/1 (accessed January16, 2013).
23 Fracturing Responsibility and Awareness of Chemicals Act. S. 1215 and H.R. 2766, 9 June 2009.
24 This provision is widely known as the earlier mentioned Halliburton loophole.
25 See www.sustainablebusiness.com/index.cfm/go/news.display/id/24384.
26 I.e. Environmental Impact Assessment Directives, Water Framework Directive, Mining Waste Directive, Directives on Emissions from Non-Road Mobile Machinery, IPCC Directive, Industrial Emissions Directive, Outdoor Machinery Noise Directive, Air Quality Directive, Environmental Liability Directive, Seveso II Directive (AEA, 2012a).
27 Recommendation 2014/70/EU.
28 See http://eenews.net/public/energywire/2012/07/16/1.
29 See http://oilprice.com/Energy/Natural-Gas/Natural-Gas-Pipeline-Bottlenecks-Lead-to-Price-Spikes-in-New-England.html.
30 See www.acer.europa.eu/Media/News/Pages/ACER-adopts-a-decision-on-the-allocation-of-costs-for-the-Gas-Interconnection-project-between-Poland-and-Lithuania.aspx.
31 See www.eustream.sk/en_media/en_news/eustreams-new-gas-interconnection-projects-are-advancing.
32 See www.pnb.pl/index.php?option=com_pnb&view=file&id=7WYSFcMtpbf DyfbnyU9&sid=b1496916&ext=21#start.
33 See www.reuters.com/article/2013/09/09/poland-energy-lng-idUSL6N0H22WR 20130909.
34 The full letter can be found on See http://democrats.naturalresources. house.gov/sites/democrats.naturalresources.house.gov/files/content/files/2012-01-04_LTR_ExportingNaturalGas.pdf (accessed October 10, 2012).
35 See http://ec.europa.eu/dgs/jrc/index.cfm?id=1410&dt_code=NWS&obj_id= 15260&ori=RSS.
36 See www.bloomberg.com/apps/news?pid=newsarchive&sid=anlPM8zJ_rE4.
37 See www.coloradoan.com/story/news/local/2014/08/07/judge-overturns-fort-collins-fracking-moratorium/13743031.
38 See www.latimes.com/nation/la-na-texas-fracking-20141108-story.html.
39 See www.reuters.com/article/2014/12/17/us-energy-fracking-newyork-id USKBN0JV29Z20141217.
40 See http://uk.reuters.com/article/2014/01/13/uk-britain-fracking-id UKBREA0C00C20140113.

41 See www.whitehouse.gov/the-press-office/2011/05/28/remarks-president-obama-and-prime-minister-tusk-poland-joint-press-confe.
42 Though more empirical work would be helpful, this EC document provides an illustration: http://europa.eu/pol/env/flipbook/en/files/environment.pdf.
43 See http://uk.reuters.com/article/2013/12/17/uk-britain-fracking-cameron-idUKBRE9BG0NX20131217.

References

AEA, 2012a. *Support to the Identification of Potential Risks for the Environment and Human Health Arising from Hydrocarbons Operations Involving Hydraulic Fracturing in Europe*. Report for DG Environment. Brussels: European Commission.

AEA, 2012b. *Climate Impact of Potential Shale Gas Production in the EU*. Final report for DG Clima. Brussels: European Commission.

Aguilera, R. F., 2014. Production costs of global conventional and unconventional petroleum. *Energy Policy* 64: 134–140.

Alvarez, R. A., Pacala, S. W., Winebrake, J. J., Chameides, W. L., Hamburg, S. P., 2012. Greater focus needed on methane leakage from natural gas infrastructure. *Proceedings of the National Academy of Sciences* 109(17): 6435–6440.

Andrews, A., Folger, P., Humphries, M., Copeland, C., Tiemann, M., Meltz. R., Brougher, C., 2009. *Unconventional Gas Shales: Development, Technology and Policy Issues*. Washington, DC: Congressional Research Service. See www.fas.org/sgp/crs/misc/R40894.pdf.

Below, A., 2013. Obstacles in energy security: an analysis of congressional and presidential framing in the United States. *Energy Policy* 62: 860–868.

Blohm, A., Peichel, J., Smith, C., Kougentakis, A., 2012. The significance of regulation and land use patterns on natural gas resource estimates in the Marcellus shale. *Energy Policy* 50: 358–369.

Boersma, T., Johnson, C., 2013. Twenty years of US experience: lessons learned for Europe. In C. Musialski *et al.* (eds), *Shale Gas in Europe: Opportunities, Risks, Challenges; A Multidisciplinary Analysis with a Focus on European Specificities*. Brussels: Claeys & Casteels Law Publishers.

Boersma, T., Khodabakhsh, C., 2014. EU engagement with shale gas. *Oil, Gas, and Energy Law Intelligence* 12(3): 1–12.

Boersma, T., Mitrova, T., Greving, G., Galkina, A., 2014. *Business As Usual: European Gas Market Functioning in Times of Turmoil and Increasing Import Dependence*. ESI policy brief 14-05. Washington, DC: The Brookings Institution.

Boudet, H., Clarke, C., Bugden, D., Maibach, E., Roser-Renouf, C., Leiserowitz, A., 2014. "Fracking" controversy and communication: using national survey data to understand public perceptions of hydraulic fracturing. *Energy Policy* 65: 57–67.

Brandt, A. R., Heath, G. A., Kort, E. A., O'Sullivan, F., Petron, G., Jordaan, S. M., Tans, P., Wilcox, J., Gopstein, A. M., Arent, D., Wofsy, S., Brown, N. J., Bradley, R., Stucky, G. D., Eardley, D., Harriss, R., 2014. Methane leaks from North American natural gas systems. *Science* 434: 733–735.

Cathles III, L., Brown, L., Taam, M., Hunter, A., 2012. A commentary on "The greenhouse-gas footprint of natural gas in shale formations" by R. W. Howarth, R. Santoro, and Anthony Ingraffea. *Climatic Change* 113(2): 525–535.

Central Statistical Office (Poland), 2012. *Wyniki Narodowego Spisu Powszechnego*

Ludnooeci i Mieszkań 2011. National census report. Warsaw: Central Statistical Office.

Davis, C., 2012. The politics of "fracking": regulating natural gas drilling practices in Colorado and Texas. *Review of Policy Research* 29: 177–191.

Deutsch, J., 2011. The good news about gas: the natural gas revolution and its consequences. *Foreign Affairs* 90(1): 82–93.

Elliot, T. R., Celia, M. N., 2012. Potential restrictions for CO_2 sequestration sites due to shale and tight gas production. *Environmental Science and Technology* 46(7): 4223–4227.

Ellsworth, W. L., Hickman, S. H., Lleons, A. L, McGarr, A., Michael, A. J., Rubinstein, J. L., 2012. Are seismicity rate changes in the midcontinent natural or manmade? *Seismological Research Letters* 83(2): 403.

European Commission, 2009. The Commission calls for proposals for €4 billion worth of energy investments. Press release. Brussels: European Commission.

European Commission, 2012. *Communication: Making the Internal Energy Market Work*. COM(2012) 663 final. Brussels: European Commission.

European Investment Bank, 2011. EIB supports Poland's transport and energy infrastructure. Press release. Warsaw/Luxembourg: European Investment Bank.

European Parliament, 2011. *Impacts of Shale Gas and Shale Oil Extraction on the Environment and on Human Health*. ENVI committee. Brussels: European Parliament. See www.europarl.europa.eu/document/activities/cont/201107/20110715ATT24183/20110715ATT24183EN.pdf.

European Union, 2009. Directive 2009/73/EC of the European Parliament and of the Council of 13 July 2009 concerning common rules for the internal market in natural gas and repealing Directive 2003/55/EC. *Official Journal of the European Union* 9/112L.

European Union, 2010a. Consolidated versions of the Treaty on the European Union and the Treaty on the Functioning of the European Union. *Official Journal of the European Union* C 83, 1-388.

European Union, 2010b. Amendment to Regulation (EC) no. 663/2009 establishing a program to aid economic recovery by granting Community financial assistance to projects in the field of energy (Regulation No 1233/2010). *Official Journal of the European Union* L 346/5.

Freyman, M., Salmon, R., 2013. *Hydraulic Fracturing and Water Stress: Growing Competitive Pressures for Water*. Research paper. Boston, MA: Ceres.

Fry, M., 2013. Urban gas drilling and distance ordinances in the Texas Barnett Shale. *Energy Policy* 62: 79–89.

Fry, M., Hoeinghaus, D. J., Ponette-González, A. G., Thompson, R., La Point, T. W., 2012. Fracking vs faucets: balancing energy needs and water sustainability at urban frontiers, viewpoint. *Environmental Science and Technology* 46(14): 7444–7445.

Hammer, R., VanBriesen, J., 2012. *In Fracking's Wake: New Rules are Needed to Protect Our Health and Environment from Contaminated Wastewater*. New York: Natural Resources Defense Council. See www.nrdc.org/energy/files/Fracking-Wastewater-FullReport.pdf.

Heath, G. A., O'Donoughue, P., Arent, D. J., Bazilian, M., 2014. Harmonization of initial estimates of shale gas life cycle greenhouse gas emissions for electric power generation. *Proceedings of the National Academy of Sciences of the United States of America* 111(31): 1–10.

Holland, A., 2011. *Examination of Possibly Induced Seismicity from Hydraulic Fracturing in the Eola Field, Garvin County, Oklahoma*. Norman, OK: Oklahoma Geological Survey. See www.ogs.ou.edu/pubsscanned/openfile/OF1_2011.pdf.

Horton, S., 2012. Disposal of hydrofracking waste fluid by injection into subsurface aquifers triggers earthquake swarm in central Arkansas with potential for damaging earthquake. *Seismological Research Letters* 83(2): 250–260.

Houser, T., Mohan, S., 2014. *Fueling Up: The Economic Implications of America's Oil and Gas Boom*. Washington, DC: Peterson Institute for International Economics.

Howarth, R., Santoro, R., Ingraffea, A., 2011. Methane and the greenhouse-gas footprint of natural gas from shale formations. *Climatic Change* 106(4): 679–690.

International Energy Agency, 2012b. *Medium-Term Gas Market Report 2012: Market Trends and Projections to 2017*. Paris: International Energy Agency.

Jackson, R. B., Rainey Pearson, B., Osborn, S. G., Warner, N. R., Vengosh, A., 2011. *Research and Policy Recommendations for Hydraulic Fracturing and Shale-Gas Extraction*. Durham, NC: Center on Global Change, Duke University.

Jenner, S., Lamadrid, A. J., 2013. Shale gas vs. coal: policy implications from environmental impact comparisons of shale gas, conventional gas, and coal on air, water, and land in the United States. *Energy Policy* 53: 442–453.

Jiang, M., Griffin, W. M., Hendrickson, C., Jaramillo, P., VanBriesen, J., Venkatesh, A. 2011. Life cycle greenhouse gas emissions of Marcellus shale gas. *Environmental Research Letters* 6(3): 1–9.

Kaiser, M. J., 2012. Profitability assessment of Haynesville shale gas wells. *Energy* 38: 315–330.

Kargbo, D. M., Wilhelm, R. G., Campbell, D. J., 2010. Natural gas plays in the Marcellus Shale: challenges and potential opportunities. *Environment, Science and Technology* 44(15): 5679–5684.

Kelsey, T. W., Shields, M., Ladlee, J. R., Ward, M., 2011. *Economic Impacts of Marcellus Shale in Pennsylvania: Employment and Income in 2009*. Williamsport, PA: Marcellus Shale Education and Training Center, State College, Pennsylvania. See www.shaletec.org/docs/EconomicImpactFINALAugust28.pdf.

Keranen, K. M., Savage, H. M., Abers, G. A., Cochran, E. S., 2013. Potentially induced earthquakes in Oklahoma, USA: Links between wastewater injection and the 2011 M_w 5.7 earthquake sequence. *Geology* 41(6): 699–702.

Kratochvíl. P., Tichy, L., 2013. EU and Russian discourse on energy relations. *Energy Policy* 56: 391–406.

Kresse, T. M., Warner, N. R., Hays, P. D., Down, A., Vengosh, A., Jackson, R. B., 2012. *Shallow Groundwater Quality and Geochemistry in the Fayetteville Shale Gas-Production Area, North-Central Arkansas, 2011*. Report 2012–5273. Reston, VA: US Geological Survey Scientific Investigations.

Kriesky, J., Goldstein, B. D., Zell, K., Beach, S., 2013. Differing opinions about natural gas drilling in two adjacent counties with different levels of drilling activity. *Energy Policy* 58: 228–236.

Kropatcheva, E., 2014. He who has the pipeline calls the tune? Russia's energy power against the background of the shale "revolutions." *Energy Policy* 66: 1–10.

Kuhn, M., Umbach, F., 2011. *Strategic Perspectives of Unconventional Gas: A*

Game Changer with Implication for the EU's Energy Security. London: EUCERS/King's College London.

Larsson, R. L., 2006. *Russia's Energy Policy: Security Dimensions and Russia's Reliability as an Energy Supplier*. Stockholm: FOI-Swedish Defense Research Agency.

Le Coq, C., Paltseva, E., 2012. Assessing gas transit risks: Russia vs. the EU. *Energy Policy* 42: 642–650.

Leteurtrois, J.-P., Duraville, J.-L., Pillet, D., Gazeau, J.-C., 2011. *Source-Rock Hydrocarbons in France*, Interim report. Paris: GCIET/CGEDD.

Majer, E. L., Baria, R., Stark, M., Oates, S., Bommer, J., Smith, B., Asanuma, H., 2007. Induced seismicity associated with enhanced geothermal systems. *Geothermics* 36(3): 185–222.

McGowan, F., 2011. Putting energy insecurity into historical context: European responses to the energy crises of the 1970s and 2000s. *Geopolitics* 16: 486–511.

McGowan, F., 2014. Regulating innovation: European responses to shale gas development. *Environmental Politics* 23(1): 41–58.

McJeon, H., Edmonds, J., Bauer, N., Clarke, L., Fisher, B., Flannery, B. P., Hilaire, J., Krey, V., Marangoni, G., Mi, R., Riahi, K., Rogner, H., Tavoni, M., 2014. Limited impact on decadal-scale climate change from increased use of natural gas. *Nature* 514: 482–485.

Medlock, K. B., 2012a. *US LNG Exports: Truth and Consequence*. Houston, TX: James A. Baker III Institute for Public Policy, Rice University. See http://bakerinstitute.org/publications/US%20LNG%20Exports%20-%20Truth%20and%20Consequence%20Final_Aug12-1.pdf

Medlock, K. B., 2012b. Modeling the implications of expanded US shale gas production. *Energy Strategy Reviews* 1(1): 33–41.

Montgomery, W. D., Baron, R., Bernstein, P., Tuladhar, S.D., Xiong, S., Yuan, M., 2012. *Macroeconomic Impacts of LNG Exports from the United States*. Washington, DC: NERA Economic Consulting.

Myers Jaffe, A., O'Sullivan, M. L., 2012. *The Geopolitics of Natural Gas*. Cambridge, MA: Geopolitics of Energy Project, Belfer Center for Science and International Affairs,

Myhrvold, N. P., Caldeira, K., 2012. Greenhouse gases, climate change and the transition from coal to low-carbon electricity. *Environmental Research Letters* 7(1): 1–8.

National Research Council, 2012. *Induced Seismicity Potential in Energy Technologies Prepublication*. Washington, DC: National Academies Press. See www.nap. edu/catalog.php?record_id=13355.

Newell, R. G., Raimi, D., 2014. Implications of shale gas development for climate change. *Environmental Science and Technology* 48(15): 8360–8368.

Nicot, J. P., Scanlon, B. R., 2012. Water use for shale-gas production in Texas, US. *Environmental Science and Technology* 46(6): 3580–3586.

Nicot, J.-P., Hebel, A. K., Ritter, S. M., Walden, S., Baier, R., Galusky, P., Beach, J., Kyle, R., Symank, L., Breton, C., 2011. *Current and Projected Water Use in the Texas Mining and Oil and Gas Industry (Draft)*. Austin, TX: Bureau of Economic Geology, University of Texas. See www.texasenvironmentallaw.com/pdfs/Report_TWDB-MiningWaterUse.pdf.

Noël, P., 2009. *A Market Between Us: Reducing the Political Cost of Europe's*

Dependence on Russian Gas. Cambridge: Electricity Policy Research Group. See www.eprg.group.cam.ac.uk/wp-content/uploads/2009/06/binder13.pdf.

North, D. C., 1991. Institutions. *Journal of Economic Perspectives* 5(1): 97–112.

Ohio Department of Natural Resources, 2012. *Preliminary Report on the Northstar 1 Class II Injection Well and the Seismic Events in the Youngstown, Ohio, Area.* Columbus, OH: Ohio Department of Natural Resources. See http://ohiodnr.com/downloads/northstar/UICreport.pdf.

Osborn, S. G., Vengosh, A., Warner, N. R., Jackson, R. B., 2011. Methane contamination of drinking water accompanying gas-well drilling and hydraulic fracturing. *Proceedings of the National Academy of Sciences* 108(20): 8172–8176.

O'Sullivan, F., Paltsev, S., 2012. Shale gas production: potential versus actual greenhouse gas emissions. *Environmental Research Letters* 7: 044030.

Paltsev, S., Jacoby, H. D., Reilly, J. M., Ejaz, Q.J., Morris, J., O'Sullivan, F., Rausch, S., Winchester, N., Kragha, O., 2011. The future of US natural gas production, use, and trade. *Energy Policy* 39(9): 5309–5321.

Pearson, I., Zeniewski, P., Gracceva, F., Zastera, P., McGlade, C., Sorrell, S., Speirs, J., Thonhauser, G., Alecu, C., Eriksson, A., Toft, P., Schuetz, M., 2012. *Unconventional Gas: Potential Energy Market Impacts in the European Union.* JRC Scientific and Policy Reports. Brussels: European Commission.

Philippe & Partners, 2011. *Final Report on Unconventional Gas in Europe.* Brussels: Philippe & Partners. See http://ec.europa.eu/energy/studies/doc/2012_unconventional_gas_in_europe.pdf.

Polish Geological Institute, 2012. *Assessment of Shale Gas and Shale Oil Resources of the Lower Paleozoic Baltic–Podlasie–Lublin Basin in Poland.* Warsaw: Polish Geological Institute, National Research Institute,

Rabe, B. G., Borick, C., 2011. *Fracking for Natural Gas: Public Opinion on State Policy Options.* Ann Arbor, MI: Center for Local, State, and Urban Policy, Gerald F. Ford School of Public Policy, University of Michigan. See http://closup.umich.edu/files/pr-16-fracking-survey.pdf.

Rahm, D., 2011. Regulating hydraulic fracturing in shale gas plays: the case of Texas. *Energy Policy* 39(5): 2974–2981.

Ratner, M., Parfomak, P. W., Luther, L., 2011. *US Natural Gas Exports: New Opportunities, Uncertain Outcomes.* Washington, DC: Congressional Research Service. See http://assets.opencrs.com/rpts/R42074_20111104.pdf.

Roth, M., 2011. Poland as a policy entrepreneur in European external energy policy: towards greater energy solidarity vis-à-vis Russia? *Geopolitics* 16(3): 600–625.

Royal Society and Royal Academy of Engineering, 2012. *Shale Gas Extraction in the UK: A Review of Hydraulic Fracturing.* London: Royal Society. See http://royalsociety.org/uploadedFiles/Royal_Society_Content/policy/projects/shale-gas/2012-06-28-Shale-gas.pdf.

Rühl, C., 2012. *Energy in 2011: Disruption and Continuity.* London: BP Statistical Review of World Energy. See www.bp. com/assets/bp_internet/globalbp/globalbp_uk_english/reports_and_publications/statistical_energy_review_2011/STAGING/local_assets/pdf/BP_Stats_2012_FINAL.pdf.

Shariq, L., 2013. Uncertainties associated with the reuse of treated hydraulic fracturing wastewater for crop irrigation. *Environmental Science and Technology* 47(6): 2435–2436.

Stephenson, E., Doukas, A., Shaw, K., 2012. Greenwashing gas: might a "transition fuel" label legitimize carbon-intensive natural gas development? *Energy Policy* 46: 452–459.

Stern, J., 2014. International gas pricing in Europe and Asia: a crisis of fundamentals. *Energy Policy* 64: 43–48.

Stevens, P., 2010. *The "Shale Gas Revolution": Hype and Reality*. London: Chatham House.

Theodori, G. L., Luloff, A. E., Willits, F. K., Burnett, D. B., 2014. Hydraulic fracturing and the management, disposal, and reuse of frac flowback waters: views from the public in the Marcellus Shale. *Energy Research and Social Science* 2: 66–74.

US Department of the Interior, 2012. *Oil and Gas: Well Stimulation, Including Hydraulic Fracturing, on Federal and Indian Lands*. Proposed rule. Washington, DC: Bureau of Land Management, Department of the Interior. See www.doi.gov/news/pressreleases/loader.cfm?csModule=security/getfile&pageid=293916.

US Energy Information Administration, 2011. *World Shale Gas Resources: An Initial Assessment of 14 Regions Outside the United States*. Washington, DC: US Department of Energy. See www.eia.gov/analysis/studies/worldshalegas/pdf/fullreport.pdf.

US Energy Information Administration, 2012a. *Annual Energy Outlook 2012*. Early release overview. Washington, DC: Energy Information Administration.

US Energy Information Administration, 2012b. *Natural Gas Wellhead Prices: Annual Overview*. Washington, DC: Energy Information Administration. See www.eia.gov/dnav/ng/ng_pri_sum_a_epg0_fwa_dmcf_a.htm.

US Energy Information Administration, 2012c. *Effect of Increased Natural Gas Exports on Domestic Energy Markets as Requested by the Office of Fossil Energy*. Washington, DC: Energy Information Administration.

US Energy Information Administration, 2014. *Annual Energy Outlook 2014*. Early release overview. Washington, DC: Energy Information Administration. See www.eia.gov/forecasts/aeo/er/early_introduction.cfm.

US Environmental Protection Agency, 2010. *Scoping Materials for Initial Design of EPA Research Study on Potential Relationships Between Hydraulic Fracturing and Drinking Water Resources*. Washington, DC: Environmental Protection Agency.

US Environmental Protection Agency, 2011a. *Investigation of Ground Contamination near Pavillion, Wyoming (Draft)*. Washington, DC: Environmental Protection Agency. See www.epa.gov/region8/superfund/wy/pavillion/EPA_ReportOnPavillion_Dec-8-2011.pdf.

US Environmental Protection Agency, 2011b. *Plan to Study the Potential Impacts of Hydraulic Fracturing on Drinking Water Resources*. Washington, DC: Environmental Protection Agency. See www.epa.gov/hfstudy/HF_Study_Plan_110211_FINAL_508.pdf.

US Environmental Protection Agency, 2012a. *Inventory of US Greenhouse Gas Emissions and Sinks: 1990–2010*. Washington, DC: Environmental Protection Agency. See www.epa.gov/climatechange/Downloads/ghgemissions/US-GHG-Inventory-2012-Main-Text.pdf.

US Environmental Protection Agency, 2012b. Action Memorandum: Request for Funding for a Removal Action at the Dimock Residential Groundwater Site […], in: III, U.E.R. (ed.), Philadelphia, PA. See www.fossil.energy.gov/programs/gasregulation/authorizations/Orders_Issued_2012/58._EPA_III.pdf.

US Environmental Protection Agency, 2012c. *Oil and Natural Gas Sector: New Source Performance Standards and National Emissions Standards for Hazardous Air Pollutants Reviews.* Washington, DC: Environmental Protection Agency. See www.gpo.gov/fdsys/pkg/FR-2012-08-16/pdf/2012-16806.pdf.

US Environmental Protection Agency, 2012d. *Study of the Potential Impacts of Hydraulic Fracturing on Drinking Water Resources.* Progress report. Washington, DC: Environmental Protection Agency. See www.epa.gov/hfstudy/pdfs/hf-report20121214.pdf.

US Government Accountability Office, 2012. *Unconventional Oil and Gas Development: Key Environmental and Public Health Requirements.* Washington, DC: Government Accountability Office. See www.gao.gov/assets/650/647782.pdf.

Vengosh, A., Jackson, R. B., Warner, N., Darrah, T. H., Kondash, A., 2014. A critical review of the risks to water resources from unconventional shale gas development and hydraulic fracturing in the United States. *Environmental Science and Technology* 48(15): 8334–8348.

Wang, J., Ryan, D., Anthony, E. J., 2011. Reducing the greenhouse gas footprint of shale gas. *Energy Policy* 39(12): 8196–8199.

Weber, J. G., 2012. The effects of a natural gas boom on employment and income in Colorado, Texas, and Wyoming. *Energy Economics* 34(5): 1580–1588.

Weijermars, R., 2011. Weighted average cost of retail gas (WACORG) highlights pricing effects in the US gas value chain: do we need wellhead price-floor regulation to bail-out the unconventional gas industry? *Energy Policy* 39(10): 6291–6300.

Wyciszkiewicz, E., Gosty ska, A., Liszczyk, D., Puka, L., Wiśniewski, B., Znojek, B., 2011. *Path to Prosperity or Road to Ruin? Shale Gas Under Political Scrutiny.* Warsaw: Polish Institute of International Affairs.

7 The European Union's future gas market structure

Tracing the US example?

Introduction

A number of studies of European natural gas markets make reference to their counterparts in the US (e.g. Creti and Villeneuve 2005; Neuhoff and Von Hirschhausen 2005; Ascari 2011; Vazquez *et al.* 2012). Generally, this market receives positive reviews in terms of being well integrated, demonstrating substantial liquidity on most of its trading hubs and high churn ratios. European institutions have been reforming European gas markets with the aim to increase competition and create one single market, yet this proves to be a lengthy and complex task that has not been completed to date (see e.g. European Commission 2007, 2012, 2014b). An example of Europe's reforms is the Gas Target Model (see Box 5.1 on page 80), which among others envisages an increase in spot market trade to realize gas prices based on gas-to-gas competition, instead of more traditional oil-indexation. Arguably, in northwestern Europe a trend towards more spot-market trade and less long-term contracts can be envisaged (Pearson *et al.* 2012; Stern 2014). Some have argued, based on an empirical test of the law of one price, that in this part of Europe gas markets in fact are reasonably well integrated (Harmsen and Jepma 2011). Renou-Maissant (2012) also reported that strong integration of gas markets in continental Europe (in a case study that used data from France, Germany, Italy, Spain, Belgium, and the United Kingdom) has been established, with work to be done in Belgium and the United Kingdom. In other studies in particular the United Kingdom and the Netherlands are been reported to have mature gas trading hubs (as an indicator for market functioning), whereas others in continental Europe fall behind (Heather 2012). Several studies make note of the slow development of spot markets in continental Europe and an increasing division within the EU, with spot market trade prevailing in the United Kingdom, the Netherlands and Belgium and long-term contracts remaining prominent in the rest of Europe (Abada and Massol 2011; Asche *et al.* 2013). Asymmetrical developments like these have been allocated to slow progress towards liberalization and competition in continental Europe (Stern 2007). Also, Spanjer (2009) suggested that the disappointing progress of competition is due to insufficient implementation of legislation as well as the lack of coordination and integration between member states.

Ascari (2011) has argued that after implementation of the third legislative package the EU will have several building blocks similar to the American model (i.e. effective unbundling of transportation and supply, regulated tariffs that are largely related to capacity and distance, and industry leading open processes of investment decisions).[1] Yet, as is elaborated in this case study, effective implementation of even existing EC legislation is far from certain. This, among other reasons, is why Makholm (2012: 172) has concluded that even an updated and expanded third package will not result in more competition or increased supply security, as competition is hindered by a wide range of institutional barriers.

In 2010 the 18th Madrid Forum invited the EC, national regulatory authorities and others to examine the interaction and interdependence of all relevant areas for network codes and to initiate a process establishing a gas market target model, comprising a vision for a future – more competitive – European gas market.[2] This call resulted in several reports about the shortcomings of the European gas system, and proposals for measures to safeguard stable and competitive gas supplies in the future (see Ascari 2011; CIEP 2011; Glachant 2011).[3] In that debate, Ascari (2011) referred to the US gas market as an example of a competitive gas market from which Europe could draw lessons, for instance that long-term contracts can be a tool to attract investments in new network capacity. Vazquez *et al.* (2012) observed that Glachant in his proposal puts more emphasis on the role of the regulatory authority to stimulate investments in new network capacity. CIEP (2011: 22) argued that the European gas market cannot be "shoehorned" into one theoretical economic model and should instead focus on attracting sufficient future supplies of natural gas, since Europe cannot afford the luxury of experimenting with its market design, because of its dependence on external resources. While the debate on the Gas Target Model is beyond the scope of this chapter, it is worth noting that fundamental differences between gas systems in Europe and the US may hinder the full embracement of the latter model. Vazquez *et al.* (2012), based on their analysis of network services coordination, concluded that the US and EU have "few common points": while the US gas system is organized in a market based setting with network services arranged in long-term contracts between producers and suppliers, in Europe network activities are preferably regulated and centrally organized at the national level.

This chapter sheds more light on the differences between the EU and US by exploring some key building blocks of the European gas system. The assessment is made by examining several components of the European natural gas system, namely available and planned infrastructure, implementation of legislation, market trade and long-term contracts, and the role of liquefied natural gas (LNG). This chapter examines this question by analyzing relevant policy documents, existing legislation and relevant academic contributions, and covers the period up to December 2014. It ends with concluding remarks and a discussion on consequences of the findings in terms of European decision-making structures.

Available and planned infrastructure

There are many reports arguing that the lack of infrastructure development has contributed to troubles in European gas markets. By now it is for instance broadly acknowledged that gas supply disruptions in 2009 (and likely also in 2006) in eastern Europe could have been mitigated if there would have been sufficient reverse flow options, adequate interconnection and gas storage facilities (European Commission 2010; Everis and Mercados EMI 2010). Others have argued that future production of unconventional natural gas in Poland could be hindered by a lack of available infrastructure, despite the desire of local policy makers to exploit its supposed gas reserves (Johnson and Boersma 2013). In 2012, the EC has established that major investments in infrastructure are still needed to safeguard security of supply. In particular the Baltic States, Finland, Malta, and the Iberian Peninsula are referred to as "gas islands" given their lack of interconnection facilities with neighboring countries. In addition single source dependency prevails in large parts of northern and eastern Europe (European Commission 2012; see also Boersma *et al.* 2014). In a broad review of the southeastern region of Europe, under auspices of the Gas Regional Initiative, progression on interconnection and capacity in that region was evaluated as "limited … given the acute lack of network integration that affects the region." (Everis and Mercados EMI 2010: 71).

In 2012 the EC published its first overview of recent investments in energy infrastructure in the member states. It had collected information on recent projects through legislation that had been adopted in 2010 and that aims to give European institutions access to more relevant data on the status quo of European investments in energy infrastructure.[4] With import dependency in Europe growing (from 48.9 percent in 2000 to 62.4 percent in 2010) realizing earlier investment estimates of €70 billion in the period up to 2020 in gas infrastructure is important.[5] Notifications from the member states demonstrate that investments in national grids were mostly minor, except for Sweden, Greece, and Poland.[6] By contrast, significant investments in cross border capacity have been reported, most notably in Germany, the Czech Republic, Italy, the Netherlands, and Greece. It is worth noting that ten member states used their option to be exempted from providing data for this analysis, making it incomplete and difficult to draw conclusions about the exact investment needs in terms of energy infrastructure.[7] The document furthermore reiterated the importance of so-called priority gas corridors, that were first mentioned in draft regulations for trans-European energy infrastructure. Looking at the corridors however, one cannot avoid the impression that almost the entire EU gas network is a priority (see Box 7.1). Adding to the complexity of attracting investments in gas infrastructure is the fact that natural gas is less competitive in Europe than for instance coal. Because of that its market share has declined, further incentivized by its current carbon emissions trading scheme (see e.g. Helm 2014).

Box 7.1 EU priority gas corridors and member states involved

- *NSI West Gas:* north/south gas interconnections in western Europe, involving Belgium, France, Germany, Ireland, Italy, Luxemburg, Malta, the Netherlands, Portugal, Spain, and the United Kingdom.
- *NSI East Gas:* north/south gas interconnections in central-eastern and southeastern Europe, involving Austria, Bulgaria, Cyprus, Czech Republic, Germany, Greece, Hungary, Italy, Poland, Romania, Slovakia, and Slovenia.
- *Southern Gas Corridor:* involving Austria, Bulgaria, Czech Republic, Cyprus, France, Germany, Hungary, Greece, Italy, Poland, Romania, Slovakia, and Slovenia.
- *Baltic Energy Market Interconnection Plan:* involving Denmark, Estonia, Finland, Germany, Latvia, Lithuania, Poland, and Sweden.

Source: Derived from draft Regulation 2011/0300 (COD), Annex 1; http://eur-lex. europa.eu/LexUriServ/LexUriServ.do?uri=COM:2011:0658:FIN:EN:PDF

It is critical to attract sufficient investments in gas transmission infrastructure, in particular when the majority of natural gas is imported (CIEP 2011). According to the EC (European Commission 2011), the required investments will not take place under business-as-usual conditions, because of problems related to permit granting, regulation and financing. As of late 2012 the EC is still under the impression that European investments in gas infrastructure shall not be sufficient to meet future demand.[8] Moreover, in light of the Ukraine crisis in 2014 it became evident that substantial parts of the European gas system continue to be underdeveloped, and vulnerable to supply disruptions (European Commission 2014b; Boersma *et al.* 2014). In a study on available gas transmission infrastructure in Europe, or the lack thereof, Correljé *et al.* (2009) argued that in case of bottlenecks "serious repercussions" have to be expected. Bottlenecks in this case are defined as a situation in which the lack of transmission capacity creates an imbalance between downstream and upstream of the pipeline. Downstream lack of capacity and uncertainty over future investments can create price spikes, while upstream there is no problem in terms of physical availability of natural gas (*ibid.*: 12). The repercussions set aside it is safe to assume that ongoing uncertainty about future markets may also have upstream consequences, such as delayed or cancelled investments in new supplies. It is worth noting that broadly accepted market mechanisms (e.g. open seasons) to determine the necessity to build additional infrastructure are not always efficient. Market signals may drive infrastructure investments, yet these signals are more likely to be correct where several market players operate, because more data likely contribute to the reliability of the reflection of future demand. Yet these market players can only be present if the infrastructure is available, creating a potential vicious circle. Hence, ensuring sufficient cross-border gas transport facilities may require a "super-regional top-down approach, as well as the involvement of European and national political authorities" (Everis and Mercados EMI 2010: 73). The EC in 2014 indicated in its energy security strategy that it aims to expand its top-down role, and several European leaders expressed their support for this, most notable then Polish Prime Minister Tusk, who pled for an energy union, including collective gas purchasing powers at the EU level.[9] In 2015 it is expected that the new EC under President Juncker reveals its plans to form this energy union, though it seems unlikely that collective purchasing of natural gas will be one feature. Given the renewed attention of the EC for energy market integration it does seem that the Ukraine crisis has given new élan to European cooperation, though arguably some of the ideas floating around seem far-fetched (Goldthau and Boersma 2014).

In the US investments in gas infrastructure are generally not considered to be reason for concern (Von Hirschhausen 2008). As elaborated in Chapter 5 several specific characteristics of the US natural gas system need to be taken into consideration. It is worth noting that the US has – next to one of the world's largest consumer markets – also been a large producer of natural gas for a number of decades. US natural gas production has hovered around 200 trillion cubic feet (5660 bcm) for several decades and only recently exploded under influence of large scale commercial extraction of gas from shale rock layers. However, long before this shale gas revolution, several hundreds of thousands of gas producing wells have been reported throughout the country.[10]

Correljé *et al.* (2009) emphasized the resemblances between the US and EU market, in terms of market structure and design. They observed that both markets seek to attract additional external supplies, and in addition most natural gas is used to generate electricity. They also stated that international conflicts of interest that can be observed in the EU frequently are similar to those at the US interstate level. Yet unlike the EU, the US has an institution to regulate the market on the federal level (i.e. FERC). This institution is hailed for facilitating the recent substantial expansion of the interstate gas-grid system (*ibid.*: 32). Arguably ACER could play a similar role in Europe, but in their analysis Correljé *et al.* (2009: 40) were rather skeptical about its mandate and proposals to increase regulatory mandates of European institutions are according to them usually considered "death-on-arrival." Admittedly the decision-making power of ACER has expanded slightly, and its decision about the Polish – Lithuanian interconnector is a good example of this.

Recent proposals of the EC for a regulation on guidelines for trans-European energy infrastructure are another example of the complex debate between member states and European institutions.[11] In 2010 the EC communication on this matter was embraced by the member states, yet the proposed regulation has been watered down and delayed since its publication in late 2011.[12] Ironically, while the proposal for a regulation aims to overhaul the existing TEN-E policy and financing framework (as has been elaborated on in Chapter 3, page XXXff.), with the compromise to draft lists of projects of common interests it shows resemblance with its (future) predecessor. For example, under TEN-E the ten-year network development plans also contain overviews of projects in three categories (i.e. projects of common interest, priority projects, and projects of European interest, the latter being primarily cross-border projects and therefore showing most resemblance with the project of common interest under the draft regulation for trans-European energy infrastructure). After all, the TEN-E policy and its three categories of energy infrastructure projects had resulted in a laundry list of projects that was so long that it eventually hindered substantial progress. Again, when everything becomes a priority, nothing really is. It is also worth noting that although adoption of the regulation has provided the EC for the first time with a structural mandate to invest in energy infrastructure, the primary responsibility for developing gas infrastructure lies still at the member state level (European Commission 2012: 9). In addition, the agreed budget for energy infrastructure, which is part of the Connecting Europe Facility, is €5.85 billion for the period of 2014–2020, comprising only a fraction of the total envelope that according to EC estimates is needed for energy infrastructure investments in that same period, and being close to half of the initial proposal of €9.1 billion.[13] As of early 2013, member states and European institutions were still arguing over what shall be projects of common interest, for member states found the plans for energy infrastructure too broad and had objections to the financial consequences. Only in 2014 the initial list of 248 projects of common interest that was published in October 2013 was reduced to 33 projects, and in October 2014 €647 million was allocated to a number of natural gas and electricity projects.[14]

The US interstate pipeline system was not built during the last decade. Although substantial expansions have been realized recently (mostly fuelled by the boom in domestic production), construction of the first interstate pipelines to transport domestically produced natural gas from the Southwest to the major consumer markets in the Midwest and Northeast already took off in the 1920s and 1930s. This was followed by investment in the central region of the country (Makholm 2012). Large-scale development of unconventional gas in several states in these regions (coal-bed methane in Wyoming, shale gas in Colorado, Texas, Louisiana, and Oklahoma) may be expected to reinforce this picture. Moreover from the 1980s onward a substantial part of the natural gas destined for markets in the Midwest has come from Canada.[15] So, in 2008 the US counted 49 different locations where natural gas could either be imported or exported through pipelines from Canada or Mexico. This marks an important difference between gas systems in the US and Europe. While in the US natural gas is consumed all over the country, in Europe the picture is different. Instead, gas consumption is mostly a western and southern European phenomenon, whereas in eastern Europe gas markets are marginal, with the exception of Romania, which has substantial domestic supplies of natural gas.[16]

Vazquez *et al.* (2012) observed that network activities in the EU are preferably regulated and centrally organized at the member state level. To them this suggested that member states' institutional power has influenced EU decision making on this matter (*ibid.*: 3). They noted that contrary to the US, where market players decide on investments and the federal FERC merely oversees that process, in the EU network planning is needed, even though a central planner in the EU would lack information in a market in which long-term contracts (still) prevail.

The above demonstrates that from an infrastructure perspective comparing the EU and the US only makes sense to a certain extent. A decades-long history of pipeline network development and several net exporting regions within the country give the US gas system also substantial different features than its European counterpart. Makholm (2012: 61) expressed skepticism about Europe's infrastructural challenges, and denied the quality of existing legislative packages in the EU, that in particular lack teeth in mandating transparency, requiring vertical separation, in uniting national regulatory rules and in powers of EU institutions in general. This is in line with concerns as expressed in the most recent EU communications and legislation regarding its internal gas market (European Commission 2010, 2011, 2012, 2014b).

Therefore, in summarizing the above, regarding available and planned infrastructure it is safe to conclude that more investments are required to meet expected demand. The budget for critical infrastructure under the Connecting Europe Facility comprises only €5.85 billion for energy infrastructure (next to natural gas also including electricity and carbon transport), which is not sufficient in view of the estimated total investments. Furthermore there are concerns whether existing regulations and institutional design can meet the challenges. There are profound concerns about the lack of coordination and integration between member states (Spanjer 2009; Correljé *et al.* 2009; Vazquez *et al.* 2012; Makholm 2012). In addition these concerns are linked to the question whether ACER, that was established to coordinate regulatory actions on the member state level, has the necessary mandate to orchestrate sufficient and timely investments in gas infrastructure (as elaborated in Chapter 3). Finally, this paragraph suggests that gas systems in Europe have been developing at substantially different paces, perchance further complicating integration (Abada and Massol 2011; Vazquez *et al.* 2012; Asche *et al.* 2013). This chapter now turns to the issue of implementation of existing legislation.

Implementation of legislation

As described in the third chapter of this book the EU has had a relatively brief but active history in energy policy making. In roughly twenty years three legislative packages were disseminated, aiming to improve market functioning by, among others, increasing transparency and further unbundling integrated gas companies. Yet in an assessment of the status quo the EC itself expressed its hesitance about the speed of progress being made:

> Today the EU is not on track to meet this deadline (completion of the internal market in 2014). Not only are member states slow in adjusting their national legislation and creating fully competitive markets with consumers' involvement, they also need to move away from, and resist the calls for, inward-looking or nationally inspired policies.
>
> (European Commission 2012)

The lack of implementation of legislation, for instance regarding vertically integrated gas companies, has also been subject to academic debate (Nowak 2010). He argued, in line with EC legislation philosophy, that unbundling is a precondition for competition in the EU gas market. Others have argued that even when existing legislation would be implemented, there are still too many loopholes to establish effective pipeline regulation, although it is not entirely clear what those loopholes would be (Makholm 2012: 61). Still, it is worth examining where member states stand with the implementation of existing legislation. Though relevant data are scarce, the EC in 2012 published an overview of both electricity and gas markets and cases where it started infringement procedures with regard to the second and third legislative package. These data were used in Table 7.1. It is worth noting that the list of member states that have not implemented existing legislation in Table 7.1 not necessarily represents the final version of this list. Member states that have initially reported that they have implemented existing legislation endured a *prima facie* check by EC officials, but it is possible that shortcomings are identified later. As Zhelyazkova (2013) observed, the EC cannot monitor everything.

Clearly much work remains to be done to implement legislation that was in fact scheduled to be implemented in March 2011. It appears that lack of implementation is a phenomenon occurring in the entire EU, and not something happening prominently in for instance eastern Europe. This is confirmed by Steunenberg and Toshkov (2009) who noted that in terms of transposition timeliness central and eastern European member states are not doing worse than the rest of Europe. They did remark that transposition is not the same as actual implementation, but that it is a prerequisite for implementation. Nevertheless according to the EC energy market development is highly divergent between member states, for instance when comparing northwestern Europe with eastern Europe (European Commission 2012: 9; see also Katz and Jepma 2012).

So what can European institutions do in cases like these, in which a substantial amount of member states fails to timely implement legislation? Formally the EC is responsible for ensuring that European law is implemented correctly.[17] In the first phase of the non-compliance procedure, called the infringement proceedings, the EC sends a reasoned opinion to the relevant member state in which it unfolds its reasons why that member state is not complying with EU law. The aim of this pre litigation phase is to offer a member state the opportunity yet to comply with the relevant legislation. If the member state fails, eventually the EC can refer the case to the European Court of Justice (ECJ) for the litigation procedure.[18]

In fact, even when cases are sent to the ECJ there are incidental cases (i.e. Belgium and Italy) in which member states do not even bother to comply with ECJ rulings and hence get convicted twice: first for non-compliance with European law and subsequently for ignoring an ECJ ruling (Börzel *et al.* 2012). The vast majority of cases, however, are solved in the early phase of infringement procedures (Panke 2007). It remains unclear why some member states settle non-compliance quickly whereas others do not (Börzel 2001; Börzel *et al.* 2012). There is some evidence that member states with

Table 7.1 Infringement procedures on the 2nd and 3rd Energy Package, as of October 29, 2012

Member state	2nd energy package (gas)	3rd energy package (gas)
Bulgaria	One case pending	Non-transposition case pending
Cyprus	No case	Non-transposition case pending
Estonia	Cases closed	Non-transposition case pending
Finland	Cases closed	Non-transposition case pending
Greece	One case pending	No case
Ireland	One case pending	Non-transposition case pending
Lithuania	Cases closed	Non-transposition case pending
Luxembourg	Cases closed	Non-transposition case pending
Poland	Two cases pending	Non-transposition case pending
Romania	One case pending	Non-transposition case pending
Slovakia	Cases closed	Non-transposition case pending
Slovenia	Case closed	Non-transposition case pending
Sweden	Cases closed	Non-transposition case pending
United Kingdom	One case pending	Non-transposition case pending

Source: data derived from EC staff working document Energy markets in the European Union in 2011, SWD(2012) 368 final, part III political influence but a lack of institutional capacity (e.g. Italy) are worse implementers than member states with limited political influence yet high capacity (e.g. Denmark) (Börzel *et al.* 2012). Also, EC and ECJ mechanisms to stimulate compliance (e.g. the transfer of financial resources and managerial know-how) may well be an effective way to reduce persistence in case of non-compliance (*ibid.*: 467). Panke (2007) observed that some cases of non-compliance demand judgments from the ECJ. These in turn create publicity and can empower proponents of compliance to put pressure on national governments. Yet it remains unclear to what extent this naming and shaming is effective in cases where so many member states fail to comply with existing legislation. Also, it is ambiguous whether eventual penalties from the ECJ are sufficient to motivate member states to comply. The laundry list of noncompliant member states as presented in Table 7.1 suggests that European institutions have a substantial amount of work left to be done. Arguably the number of member states that is compliant increases over time, but one key concern of the EC continues to be the lack of a collaborative approach and more regional integration (European Commission 2014b: 3).

Market trade and long-term contracts

Through its liberalization process, European institutions intended to create one European energy market. Yet after three legislative packages, barriers appear to remain intact that hinder an increase in trade of natural gas in most parts of Europe. Cavaliere (2007) concluded that gas trade has been hindered by asymmetric implementation of legislative packages aimed at liberalization, but also by a lack of interconnection capacity and the exemption of transit pipelines from regulated third party access (see also Spanjer 2009).

In 2006 ERGEG launched the Gas Regional Initiative (GRI), with the goal to speed up the integration of Europe's national gas markets.[19] As an interim step towards creating one single European gas market GRI created three regional markets with the aim to facilitate a top-down push forward, by addressing competition distortions, such as lack of transparency, inefficiencies in balancing regimes and the lack of market integration. One of the main issues to address in the GRI was related to the further development of hub-based trading across Europe (Everis and Mercados EMI 2010).

In 2010, Everis and Mercados EMI observed that while in the north-western region the issue of hub trade seems to have been picked up, there was less progress to report in the south–southeast region. In the south region the only achievement was an analysis of the situation of gas hubs in the region. The authors therefore concluded that a top-down approach is required to establish operational minimum requirements at the European level (*ibid.*: 86). Cavaliere (2007: 39) concluded that the case of Italy shows that *ex post* regulation to foster gas trade was not sufficient to remove the

Figure 7.1 Three regions of the Gas Regional Initiative

Note: France is part of both north-west and south clusters

Source: data derived from www.energy-regulators.eu/portal/page/portal/EER_HOME/EER_
    ACTIVITIES/EER_INITIATIVES/GRI

bottlenecks, because some of these barriers are outside the national network (e.g. pipelines in other countries being owned by the incumbent gas company) and likely require *ex ante* regulation at the European level. Abada and Massol (2011) observed that spot markets in continental Europe are developing slowly and suggested that retailers in Europe still have an interest in engaging in long-term trade. It is worth noting that while in northwestern Europe markets are moving away from long-term contracts (see also Harmsen and Jepma 2011; Renou-Maissant 2012; Stern 2014), in eastern Europe the upstream market structure is more concentrated and long-term contracts prevail (Abada and Massol 2011). This distinction is confirmed by Asche *et al.* (2013) who noted that spot market trading has been established in the United Kingdom, the Netherlands, and Belgium, but that long-term contracts remain prominent in continental Europe. Their results also suggested that gas prices in Europe are still determined by oil prices, but that it is unclear whether this is due to insufficiently deregulated gas markets.

Currently gas trade in northwestern Europe continues to increase, with not just the British National Balancing Point (that has a record of being a mature trading hub for over a decade) but also the Dutch Title Transfer Facility showing rapid growth. In late 2012, the EC published data indicating a 27 percent increase of gas trade from 2009 to 2010.[20] The development of gas trade in the Netherlands has been attributed to a number of factors, including the 2009 decision of Gas Transport Services to allow for quality conversion at TTF, a "real time" balancing regime that proved to work well and increased available market information. Also, an interest for the Dutch gas producer GasTerra to monetize its assets by selling gas on the market played an important role, and finally that TTF since May 2012 offers the first cross-border market coupling scheme, combining transport services of the Netherlands and northern Germany (Heather 2012).[21] His study also concluded that Europe has a long road ahead towards the creation of one single gas market, and that it is likely to expect a limited number of liquid and high volume trading hubs, together with a number of national hubs that show price correlation but not so much trade (*ibid.*: 44). Stern (2014) concluded that by late 2012 sellers of natural gas from Norway and the Netherlands appear to have shifted substantially to hub-based prices, sellers from Russia and Algeria had not, notwithstanding rebates and price negotiations. In addition, the transition to more hub-based pricing in continental Europe should not be expected to come with the termination of long-term contracts. Noël (in Kalicki and Goldwyn 2013) observed that the only country in central and eastern Europe where declining spot-market prices have really put pressure on wholesale prices is the Czech Republic, due to its connection to the German market, and to a less extent Austria and Slovakia.

It is worth considering the debate about the role of long-term contracts in European gas markets, which, as touched upon earlier, also features in the discussion about the Gas Target Model. Neuhoff and Von Hirschhausen (2005) noted that European long-term gas contracts used to be oil-indexed to protect the buyers of natural gas against prices higher than those of competing fuels. Thus these contracts functioned as a risk sharing device: the buyer of gas bears a volume risk (a minimum amount of natural gas to be purchased) and the seller has a price risk (in case prices for natural gas rise above the fixed price as agreed in the contract). Creti and Villeneuve (2005) observed that because these long-term contracts are inflexible in terms of demand and supply fluctuations, they usually contain clauses (e.g. a price floor with options to raise tariffs under predefined conditions, or a renegotiation of terms at predetermined intervals). It is worth noting that with the proceeding of liberalization of the EU gas markets, the duration of long-term contracts has diminished, from a 25-years period in 1980 to a 15-year period in 2000 (Von Hirschhausen and Neumann 2008). Finon and Roques (2008) noted, in a study on investments in nuclear power plants, that long-term contracts can have positive effects on investment. De Hautecloque and Glachant (2009) also mentioned that long-term contracts ensure investments and reliability (at a hidden cost for society), but also pointed at some of their downsides, most notably the possibility of price restraints and foreclosure (when a significant share of demand is tied up in long-term contracts this may become an entry barrier for third parties).

In 2007 the EC reported in its sector inquiry that long-term contracts in European gas markets were one of the persistent barriers for new entrants into the upstream market. If these contracts were concluded by dominant firms and foreclosed the market, they may breach with competition rules, unless there were countervailing efficiencies benefiting consumers (European Commission 2007: 10). Also, the EC concluded that the aforementioned status quo "does not as such put into question existing and future upstream contracts." This seems in line with an earlier observation of Creti and Villeneuve (2005) that the EC appeared to let the market decide about the future of long-term supply contracts and have them coexist with spot market trade.

Stern (2009) observed that decoupling has become "inevitable" because of the gas supply surplus. Pearson *et al.* (2012) reported that this abundance in natural has been further increased by the large scale extraction of unconventional natural gas in the US. According to EC figures from 2011 the difference between spot market and long-term contract gas prices further increased due to increases in oil-indexed prices (European Commission 2012).[22] These developments have made the difference between spot-market prices and those in long-term contracts so large that several utility companies had to renegotiate their contract terms with for instance Russian supplier Gazprom.[23] Figure 7.2 shows the development of prices of natural gas in €/MWh at four important global benchmarks (namely Henry Hub in the US, British hub NBP, German border prices, and the largest LNG markets in Asia) during the last decade.

Figure 7.2 shows that in the second half of 2008 prices at the trading hubs Henry Hub and NBP dropped sharply. This is attributed to the earlier mentioned gas surplus (e.g. Stern 2009), in combination with the economic downturn. The data also show that prices at the German border and Asian LNG prices continued to rise in the second half of 2008 and dropped significantly later. Although German border tariffs are not an ideal reflection of long-term contract tariffs in Europe, they are used as an indicator. Hence, following these data, in the second part of 2008 and in early 2009, natural gas under long-term contracts in Europe was almost twice as expensive as spot-market traded gas at the British NBP. Stern (2014) explained that this situation put pressure on producers of gas to lower tariffs under existing

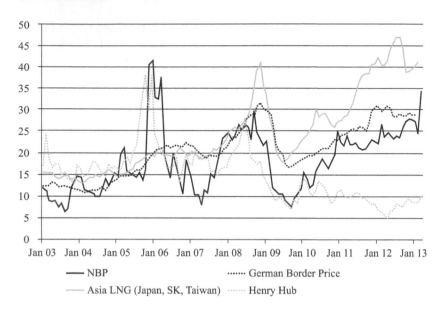

Figure 7.2 Price developments at Henry Hub, NBP, German border and Asia LNG

Source: data derived from www.customs.go.jp/english, www.bafa.de/bafa/de and
        www.eciaonline.org/eiastandards (with special thanks to Thijs van Hittersum)

long-term contracts, and in addition that new long-term contracts are since increasingly indexed to prices on liquid gas trading hubs in northwestern Europe, such as NBP, also TTF in the Netherlands or Zeebrugge in Belgium. The data also show that since mid-2009 prices at the German border and British NBP have converged. This may confirm that existing long-term contracts have been renegotiated and that the prices in renewed contracts are more in line with hub prices, such as NBP. This would not necessarily mark the end of long-term contracts in Europe, for earlier mentioned advantages of those contracts in terms of security of demand and supply (e.g. De Hautecloque and Glachant 2009) are still in place. Rather, contract conditions may increasingly be altered and move towards hub-based prices. More empirical work would be helpful here.

Stern and Rogers (2011) have concluded that there is no commercially viable alternative to hub-based pricing in the European gas market. They refer to a process that has taken place in the US in the 1980s and 1990s and note that the actual shift to hub-pricing mechanisms was in fact completed within a few years (*ibid.*: 34). They also noted that revising existing contracts is painful, with difficult negotiations and litigation in foresight. In addition, in contrast to the US where the majority of those involved fall under the same political and legal jurisdiction, continental Europe is highly diverse, as are its major suppliers. Boussena and Locatelli (2013) observed an increased difference of opinion between Europe and Russia and attributed this to rules and standards that are based on different values and beliefs.

It is impossible to assess what exactly the future of oil-indexed gas contracts in the EU will be. The EC has reported a significant fall in the share of oil-indexed gas contracts in 2010 (from 68 percent of natural gas consumption to 59 percent in the year after), due to an increase in spot purchased gas (27 percent of gas consumption 2009 and 37 percent in 2010).[24] Anecdotal evidence from the Netherlands suggested that roughly 60 percent of gas trade in 2012 has been spot-based and that the process towards full hub-based trading is expected to be completed within five years. This means that new contract structures and formulas will become common sense, based on year and month terms instead of long-term, all lined to spot-trade prices. However, there is strong evidence that in continental Europe this transition cannot be expected within the next five years, if at all (Abada and Massol 2011; Asche *et al.* 2013). Many unknowns make it difficult to predict the future of oil-indexed contracts in continental Europe, such as future supplies in general, LNG demand and supplies elsewhere in the world, the design of future carbon policies and the duration of the economic downturn.

It is complex to assess what the previous developments indicate in terms of EU decision making. The EC finds it important that hub-trading is further developed in the member states, because it is under the impression that member states with well-developed gas hubs have benefitted from greater price stability, but also because prices of imported gas under long-term contracts have been lower in these member states (European Commission 2012).[25] It is important to keep in mind though that prices based on hub-trading are not per definition lower than those based on oil prices: it depends on the market conditions. At the same time the development of spot markets across the EU has developed asynchronous, with markets in northwestern Europe being reasonably well developed whereas markets in continental Europe fall behind (Harmsen and Jepma 2011; Abada and Massol 2011; Renou-Maissant 2012; Asche *et al.* 2013). Currently there are no indications that European institutions desire to prescribe the markets in what fashion natural gas should be contracted or traded, and hence long-term contracts and spot-market trade continue to coexist in the EU, following the interplay of preferences of producers and suppliers.

The role of liquefied natural gas

The share of liquefied natural gas (LNG) in the EU has risen steadily from 10 percent twenty years ago to almost 20 percent in 2011 (European Commission 2012). It is worth noting that in 2006 only a few countries appeared interested in LNG facilities, as Table 7.2 demonstrates.

Table 7.2 Status of European Union regasification plants by country, 2006

Member state	On-stream	Under construction	Planned
Belgium	1	1	0
Cyprus	0	0	1
France	2	1	3
Germany	0	0	1
Greece	1	1	2
Ireland	0	0	1
Italy	1	2	13
Latvia	0	0	1
Netherlands	0	0	3
Poland	0	0	1
Portugal	1	0	1
Spain	5	4	5
Sweden	0	0	1
United Kingdom	1	3	6
Total	12	12	39

Source: IEFE data derived from Dorigoni and Portatadino (2008)

The role of LNG is part of several ongoing debates. First, while LNG's contribution to diversification of gas supplies in Europe is undisputed, Dorigoni and Portatadino (2008) questioned whether an increase in LNG would enhance competition on that market, since their data showed that in 2006 in Italy 73 percent of LNG capacity was operated by incumbent energy companies. Paraja (2010), however, noted that LNG did increase the level of competition in the Spanish market, with incumbent companies owning less than 45 percent of natural gas in the market.

Second, it is unclear what exactly the role of LNG in Europe will be. Though the costs of LNG are generally higher than those of natural gas that is transported by pipeline, Dorigoni *et al.* (2010) suggested that LNG from countries like Libya, Algeria and Qatar can in Europe effectively compete with Russian gas. Lochner and Bothe (2009), to the contrary, expected the share of LNG in Europe to decline up to 2030, since Europe's geographical location allows it to import large quantities of natural gas by pipeline at moderate costs. Kumar *et al.* (2011) predicted that the share of LNG in Europe may rise to 25 percent in the long term, although it is unclear when that would exactly be. Boersma *et al.* (2014) concluded that the share of LNG in Europe likely increases in the future, mostly due to dwindling domestic production of natural gas. Too often though policy makers (in light of the Ukraine crisis) mistake LNG imports as a substitute for Russian natural gas, which is believed to be at least 30 percent cheaper based on marginal costs. What adds to these uncertainties is the lack of clarity how much additional supplies will come to the market in the coming years, and how large natural demand will be. A part of this debate is the question whether the US is going to export significant volumes of natural gas. At this stage it is not clear how much natural gas from the US will be able to compete directly in Europe, because it is generally more expensive than conventional natural gas that reaches Europe by pipeline from for instance Russia and Norway (Paltsev *et al.* 2011). Boersma *et al.* (2014) do expect US LNG to be competitive in the more liquid markets of northwestern Europe, particularly the United Kingdom, the Netherlands, and Belgium. In addition, large quantities of LNG from the US may displace LNG from for instance the Middle East and Nigeria to Europe (Henderson 2012). As more LNG supplies are expected to enter the market in the second half of this decade and onward, from places as diverse as Australia, Mozambique, Papua-New Guinea, Russia, the US, and Canada, it is likely that LNG spot market prices will come down, possibly making LNG more competitive in Europe.

Industry data, however, suggested that in the short-term the center of gravity for LNG trade is increasingly in Asia. As of late 2012, Asia accounted for 71 percent of global LNG demand, in comparison to 64 percent in 2011 (GIIGNL 2012). This increase is attributed to Japan's closure of nuclear power plants in the post-Fukushima era, and also continued growth in upcoming LNG markets such as China and India. Europe's share in global LNG trade, to the contrary, fell from 27 percent in 2011 to 20 percent in 2012, a decline that is attributed by the economic downturn and subsequent decreasing demand. The (generally) higher prices for natural gas in Asia (as depicted earlier in Figure 7.2) and substantial room for growth in natural gas consumption in countries like India and China may further increase this trend.

Finally, LNG also features in a broader debate about natural gas and its role as a "transition fuel" to a low carbon economy. It is debated to what extent unconventional natural gas can be considered a clean fossil fuel, because the production process of LNG, that adds liquefaction, tanker transport and regasification to the life cycle of natural gas, also results in an increase in carbon emissions (Stephenson *et al.* 2012). More research is required into the carbon footprint of both unconventional gas and LNG.

As Table 7.3 demonstrates, by 2012 the number of regasification plants in Europe had almost doubled since 2006 and four more plants are under construction. It is worth considering the position of the Iberian Peninsula, in particular Spain, where LNG features dominantly in the national energy mix, comprising 76 percent of imported natural gas in 2010.[26] Spain imports natural gas from fourteen different countries, and its most prominent suppliers are Algeria (33 percent), Nigeria (20 percent), and Qatar (16 percent). So, the Iberian Peninsula has developed into an energy island and has one of the lowest levels of interconnection in Europe. Of the total entry capacity of the Spanish gas market, 6 LNG regasification terminals provide 49 bcm, while 20 bcm comes through two pipelines with respectively

Table 7.3 Status of European Union regasification plants by country, as of June 2012

Member state	On-stream	Under	Planned construction	Suspended
Albania	0	0	1	0
Belgium	1	0	0	0
Croatia	0	0	1	0
Cyprus	0	0	0	1
France	3	1	0	1
Germany	0	0	0	1
Greece	1	0	0	0
Ireland	0	0	1	0
Italy	2	1	5	0
Netherlands	1	0	0	0
Poland	0	1	0	0
Portugal	1	0	0	0
Spain	6	1	1	0
Sweden	1	0	0	0
United Kingdom	4	0	0	1
Total	20	4	9	4

Source: data based on available information on www.globallnginfo.com (accessed December 3, 2012)

Morocco and Algeria. With European neighbor France there is only a 2.5 bcm interconnector that allows gas to flow from France to Spain, despite repetitive calls from the EC that this situation should be improved (European Commission 2012). There are plans for an extra gas interconnection facility in Catalonia. Meanwhile Spain has been investing to increase its storage capacity fourfold up to 2016, in order to deal with possible supply disruptions in LNG delivering countries and bad weather conditions that complicate landing of LNG ships (Paraja 2010). It is not clear to what extent Spanish dependence on LNG supplies is a deliberate choice of Spanish policy makers following the mediocre connection to the European gas grid, or whether LNG is just more competitive given the distance between upstream activities supplying the EU and the Iberian Peninsula. In general LNG is favorable to pipeline transmission of natural gas for distances larger than 1500 kilometers offshore or 3000 kilometers onshore (Von Hirschhausen *et al.* 2008).

In 2008 the EC commissioned a study whether it should play a more active role in the LNG supply chain, which consists of upstream natural gas production and distribution, liquefaction, shipping, regasification, and finally storage, transmission and distribution. The results showed that there was limited need for an EU Action Plan for LNG for several reasons (*ibid.*). First, from an economic point of view the authors found no reasons that would justify EU action; increased LNG trade brings benefits in terms of security of supply, though increased global competition for natural gas can be a risk as well. Future low-carbon scenarios, however, predict a decrease of natural gas usage from 2030 onward since it is not expected to be competitive once coal can be combined with sequestration technologies. Also, most upstream activities of LNG are outside the sphere of influence of the EU, besides strategic partnerships it already makes. Second, from a regulatory perspective the authors examined mostly benefits of LNG in terms of security of supply. They concluded that security of gas supply (LNG, pipeline and domestic production) is not in danger in Europe, provided that the internal market works (*ibid.*: 5).

The 2014 Ukraine crisis contained some valuable lessons regarding the possible role of LNG in terms of European energy security. At the onset policy makers, and several scholars, on both sides of the Atlantic Ocean cried out for more LNG import capacity in Europe, so that the continent could move away from Russian natural gas supplies (see Goldthau and Boersma 2014). These outcries seemed to overlook the fact that at the time Europe in fact had close to 200 bcm regasification capacity in place, ready to be utilized at any given time. The commercial reality, however, was that LNG was (and is) relatively expensive compared to pipeline gas, albeit from Norway, or Russia. With natural gas demand in general being in decline (as discussed, further incentivized by a dysfunctional emissions trading scheme and increased energy efficiency), there was no incentive to attract natural gas in the form of LNG, and the average utilization rate of LNG terminals across Europe hovered around 20 percent (Boersma *et al.* 2014). Thus, it was clearly demonstrated that without a willingness to pay a substantial premium for LNG, most supplies would in fact be sold elsewhere, predominantly in Asia, and not in Europe. Unsurprisingly, political preference was not (and is unlikely to become) part of the commercial lexicon. The fact that LNG imports at this point in time are not an attractive commercial proposition is underlined by the observation that as of August 2014 seven planned LNG regasification terminals (in Italy, Spain, Cyprus, the UK, the Netherlands, France, and Germany) have been suspended or cancelled.[27] Market forces may, however, alter the position of LNG in the European market. Ruester (in Glachant *et al.* 2013) observed that around 20 percent of global LNG trade is done at the spot-market and in short-term contracts. This, together with changing contracting formulas with more flexibility and new pricing mechanisms (in the United States in particular), make it likely that at least a part of the global LNG market becomes flexible, and supply and demand patterns will determine where the center of short-term LNG trade will be (for an in-depth study on future LNG contracts, see Hartley 2013).

New LNG terminals that will start operations in 2015 are located in Poland and Lithuania. As discussed earlier, the economics of both these terminals have to be questioned. First, significant public funds were required to construct the terminals, and more importantly, natural gas purchases in the form of LNG, from Qatar and Norway respectively, are going to be more expensive than Russian natural gas. In central and eastern Europe however, these cases demonstrate a willingness to pay a premium for non-Russian natural gas. The long-term effects of non-market based interventions like these deserve more empirical analysis. For now it seems fair to conclude that the costs of these terminals, which provide access to alternative supplies, could be seen as pricy insurance policies. In central and eastern Europe LNG facilities may bring benefits, but inverting pipeline capacity to facilitate imports of natural gas from western Europe may be equally beneficial to these single source dependent member states. As Table 7.4 suggests, these countries currently lack sufficient transport and interconnection capacity, and without substantial investments therefore parts of Europe, most notably central and eastern Europe, are unlikely to benefit from the natural gas that is available in the EU, for natural gas cannot physically flow there. As an illustration of the modest pace at which member states in central and eastern Europe address what are often called "security" concerns, Table 7.5 demonstrates an evaluation from May 2014, which the EC published in light of the Ukraine crisis.

Discussion

This analysis of several important features of the EU gas system has shown the following. In terms of available gas transmission infrastructure, substantial investments are required to complete the internal gas market. Also, under business as usual conditions these investments are not expected to happen in time (European Commission 2012). The Ukraine crisis in 2014, demonstrated once more that Europe does not have an integrated gas market, and that substantial parts of the continent are not properly harnessed against a supply disruption, even though important and significant progress has been made between 2009 and 2014. Countries like the Czech Republic and Poland are currently better off, and as their examples show, market integration actually works. It is also important to keep in mind that during the Ukraine crisis Russia did not physically end supplies to Europe, and even went at length to emphasize that it did not intend to cut supplies either. This confirms the mutual dependence that has been a feature of EU–Russian energy relations for decades, even though at times rhetoric and opportunistic policy may take over.

Contrary to the US, in the EU there is lack of an institute to orchestrate the most urgently required investments, as infrastructure development remains a primary responsibility of the member states. The expected decision regarding critical infrastructure investments under the Connecting Europe Facility does not fundamentally alter that picture. While, for example, Correljé *et al.* (2009) have argued that the mandate of ACER should be expanded to orchestrate these investments, this chapter suggests that would not be a panacea. That is because, next to lacking a supranational

Table 7.4 Infrastructural bottlenecks in central and eastern Europe

Member state	Infrastructural bottleneck
Austria	No reverse flow options on the HAG pipeline, connecting Austria with Hungary.
Bulgaria	A gas country under construction: Domestic infrastructure is not connected to large transit pipelines, there are limited interconnection and reverse flow options with Greece and Romania, and there are hardly storage facilities available.
Czech Republic	The country has successfully invested in the Gazelle pipeline, which links it to Nord Stream in Germany. More investments are needed in reverse flow facilities on existing pipelines, as are investments in storage facilities.
Estonia	In short, Estonia is an "energy island" that requires significant investments.
Germany	Significant bottlenecks remain in terms of reverse flow options at the Polish border.
Hungary	Physical interconnection with Slovakia is lacking, as are reverse flow options with Romania. The connection with Austria is one-directional.
Latvia	Latvia is also typified as an "energy island," where significant investments are needed to develop the gas system.
Lithuania	Like its Baltic neighbors, Lithuania is an "energy island" though an interconnector with Poland is under study.
Poland	Poland has made significant investments in recent years to develop its market. The domestic grid system needs substantial investment, as should regulated market policy and the dominance of the incumbent PGNiG be reduced. Also the Yamal pipeline lacks bi-directional facilities to allow reverse flow from Germany.
Romania	According to the European Commission much work needs to be done, despite substantial domestic production of natural gas. Domestic infrastructure needs to be connected to large transit pipelines. Also, investments should be made in an interconnector with Bulgaria, as in reverse flow options with Hungary.
Slovakia	An interconnector with Hungary has been planned, not built.

Source: data derived from EC staff working document SWD(2012) 368 final, part II, accompanying COM(2012) 663 final

institute that guards over infrastructure investments, another important difference with the US is that in the US consumers are spread out across the continent, whereas in Europe the east and southeast are undeveloped, in terms of gas consumption. Several studies have confirmed the asynchronous development of parts of the European gas system (e.g. Abada and Massol 2011; Asche *et al.* 2013). A repeatedly addressed problem in the EU is the lack of coordination and integration (e.g. Spanjer 2009; Correljé *et al.* 2009; Vazquez *et al.* 2012; Makholm 2012). It is also worth noting

Table 7.5 Planned infrastructure projects in central and eastern Europe and their estimated completion time

Project	Details	Finished by
Klaipeda–Kiemena pipeline upgrade and Latvia	Capacity enhancement of the interconnector between Lithuania	2017
EL–BG interconnector	New interconnector between Greece and Bulgaria to support diversification and deliver Shah Deniz gas in Bulgaria	2016
EL–BG reverse flow	Permanent reverse flow on the existing interconnector between Greece and Bulgaria	2014
BG storage upgrade	Increase storage capacity in Chiren, Bulgaria	2017
HU–HR reverse flow	Reverse flow enabling gas flow from Croatia to Hungary	2015
HU–RO reverse flow	Reverse flow enabling gas flow from Romania to Hungary	2016
BG–RS interconnector	New interconnector between Bulgaria and Serbia	2016
SK–HU interconnector	New bi-directional pipeline between Slovakia and Hungary, currently under construction	2015
PL–LT interconnector	New bidirectional pipeline, ending isolation of Baltic states	2019
FI–EE interconnector	New bidirectional pipeline between Finland and Estonia	2019
LV–LT interconnector	Upgrade of existing interconnector between Lithuania and Latvia	2020
PL–CZ interconnector	New bidirectional pipeline between Poland and Czech Republic	2019
PL–SK interconnector	New bidirectional pipeline between Poland and Slovakia	2019
PL: 3 internal pipelines and compressor station	Internal reinforcements required to link Baltics with region south of Poland	2016–2018
BG: internal system	Rehabilitation and expansion of transport system needed for regional integration	2017 (to be confirmed)
RO: internal system and reverse flow to UA	Integration of Romanian transit and transmission system, and reverse flow to Ukraine	to be decided

Source: based on European Commission (2014b)

that the US market did not develop overnight, but instead went through several decades of institutional history before it became the integrated gas system it is today.

In terms of implementation of existing legislation, European member states and institutions have another long road ahead. As of late 2012, infringement procedures are pending against fourteen member states, and that may not be the final number. Since then progress has been made, but anecdotal evidence suggests that there is still a handful of countries that have to implement existing legislation. Though the available literature suggests that naming and shaming works in the majority of cases (e.g. Panke 2007), it is not clear whether that will also be the case when so many member states are noncompliant, as is the case here. Also, there is no clear evidence what the reasons are for rapid implementation of legislation, or the lack thereof.

Market trade is still hindered, despite several legislative packages. Reasons that have been mentioned for this are the lack of tradable gas (liquidity), inefficient use of transmission capacity, strategic withholding of transmission capacity, and asymmetric implementation of legislation across member states, lack of interconnection capacity, contractual congestion and the many exemptions of transit pipelines from so-called third-party access. As of 2012, several studies suggested that markets in northwestern Europe are reasonably well integrated (Harmsen and Jepma 2011; Renou-Maissant 2012; Heather 2012), whereas in other parts of Europe long-term contracts prevail (Abada and Massol 2011; Asche *et al.* 2013). Currently it is unclear what exactly the role of long-term contracts in Europe will be. The EC (European Commission 2012) has reported an increase in spot-market trade and appears to embrace that development. Oversupply on the markets since mid-2008 have further increased the difference between spot-market prices and those in long-term contracts, sometimes resulting in renegotiation of contracts by suppliers and producers. Some studies have suggested that the phasing out of long-term contracts is unavoidable (Stern 2009; Stern and Rogers 2011), whereas others also mentioned that long-term contracts can have positive effects on investment (Finon and Roques 2008; De Hautecloque and Glachant 2009). It appears that the EC intends to let long-term contracts and spot-market trade coexist, but that the conditions for long-term contracts in Europe may increasingly shift away from oil-indexation to for instance hub-based indexation, more in line with gas prices on for instance British NBP or the Dutch TTF (see also Stern 2014).

LNG currently contributes to European energy security in terms of diversification and recent investments in LNG regasification terminals confirm this trend, although scholars do not agree about the future role of LNG in Europe, with some suggesting an increase of LNG in the next decades (Dorigoni *et al.* 2010; Kumar *et al.* 2011; Boersma *et al.* 2014), and others predicting exactly the opposite (Lochner and Bothe 2009). Industry data suggested that as of late 2012, the center of gravity for global LNG demand has further shifted towards Asia, whereas Europe's share has fallen from 27 percent in 2011 to 20 percent in 2012 (GIIGNL 2012). One major determining factor will be what overall gas demand in Europe will be. In recent years, coal has become more competitive than natural gas for electricity generation, and without substantial fixes in terms of carbon prices, this situation is unlikely to change. Increased energy efficiency also drives down natural gas demand. Thus, the paradoxical situation is that Europe has ambitious renewables and carbon reduction targets, but at the same time its dysfunctional carbon pricing scheme incentivizes an increased usage of coal at the expense of natural gas (e.g. Helm 2014). This combined with geopolitical concerns and the urge to drive down gas demand in order to reduce dependence on Russia following the Ukraine crisis, has put natural gas in an unfavorable position. Smith (2013) concluded that EU gas demand is unlikely to rise before 2020, and may remain close to current levels out to 2030. It is also worth noting that member states can only benefit from LNG supplies in a functioning internal gas system. As this chapter has argued, this is currently not always the case, certainly not in central and eastern Europe.

Notes

1   For a brief overview of this EU third legislative package, see for instance http://ec.europa.eu/energy/gas_electricity/legislation/doc/20110302_entry_into_force_third_package.pdf (accessed November 26, 2012).
2   The European Gas Regulatory Forum (Madrid Forum) has gathered once or twice since 1999 and was set up to discuss issues regarding the creation of the internal gas market; see http://ec.europa.eu/energy/gas_electricity/gas/forum_gas_madrid_en.htm.
3   In total five proposals for a gas target model have been submitted to ACER. The two proposals that are not addressed in this book are from Moselle and White (2011) and Frontier Economics (2011). Since this book does not get into the details of this ongoing debate these contributions are only mentioned here.
4   Regulation 617/2010; see http://eur-lex.europa.eu/LexUriServ/LexUriServ.do?uri=OJ:L:2010:180:0007:0014:EN:PDF.
5   For EU import dependency statistics, see for instance http://epp.eurostat.ec.europa.eu/statistics_explained/index.php?title=File:Energy_dependency_rate,_EU-27,_2000-2010_%28%25_of_net_imports_in_gross_inland_consumption_and_bunkers,_based_on_tonnes_of_oil_equivalent%29.png&filetimestamp=20121012131838.
6   Data used in this section come from a working document by Commission staff on investment projects in energy infrastructure: SWD(2012) 367, p. 15ff.
7   Regulation 617/2010, art. 3, sub 2. The ten member states referred to are Austria, Germany, France, Ireland, Lithuania, Latvia, the Netherlands, Spain, Sweden, and the United Kingdom.
8   EC staff working document Investment projects in energy infrastructure SWD(2012) 367, p. 17.
9   See   www.economist.com/blogs/easternapproaches/2014/04/donald-tusks-energy-union.
10 For data, see for instance www.eia.gov/dnav/ng/hist/na1170_nus_8a.htm.
11 2011/0300 (COD).
12 COM(2010) 677 final.

13 The total envelope for energy infrastructure was estimated to be €200 billion up to 2020, of which €70 billion for gas transmission infrastructure (European Commission, 2011).
14 See http://ec.europa.eu/energy/infrastructure/pci/pci_en.htm.
15 For the full US EIA report, visit www.eia.gov/pub/oil_gas/natural_gas/analysis_publications/ngpipeline/index.html (accessed December 2, 2012).
16 See http://ec.europa.eu/energy/energy_policy/doc/factsheets/mix/mix_ro_en.pdf.
17 See article 285ff. of the Treaty of the Functioning of the European Union (TFEU).
18 See also http://ec.europa.eu/eu_law/infringements/infringements_en.htm.
19 See www.energy-regulators.eu/portal/page/portal/EER_HOME/EER_ACTIVITIES/EER_INITIATIVES/GRI.
20 Commission staff working document, SWD(2012) 368 final, part I.
21 This is important because it allows traders to supply or take both high- and low-caloric natural gas to and from the system, necessary because the large Groningen field in the Netherlands contains low-caloric while most imported gas is high-caloric.
22 Commission staff working document SWD(2012) 368 final, part I, p. 23.
23 See for instance www.fitchratings.com/gws/en/fitchwire/fitchwirearticle/EUUtilities-May?pr_id=769294
24 EC staff working document SWD(2012) 368 final, part I.
25 *Ibid.*, p. 31.
26 SWD(2012) 368 final, part II, p. 186.
27 See www.globallnginfo.com/Introduction.htm#g (accessed November 25, 2014).

References

Abada, I., Massol, O., 2011. Security of supply and retail competition in the European gas market: some model-based insights. *Energy Policy* 39(7): 4077–4088.

Ascari, S., 2011. *An American Model For the EU Gas Market?* Working paper RSCAS 2011/39. San Domenico di Fiesole: Robert Schuman Center for Advanced Studies, European University Institute.

Asche, F., Misund, B., Sikveland, M., 2013. The relationship between spot and contract gas prices in Europe. *Energy Economics* 38: 212–217.

Boersma, T., Mitrova, T., Greving, G., Galkina, A., 2014. *Business As Usual: European Gas Market Functioning in Times of Turmoil and Increasing Import Dependence.* ESI policy brief 14-05. Washington, DC: The Brookings Institution.

Börzel, T. A., 2001. Non-compliance in the European Union: pathology or statistical artifact? *Journal of European Public Policy* 8(5): 803–824.

Börzel, T. A., Hofmann, T., Panke, D., 2012. Caving in or sitting it out? Longitudinal patterns of non-compliance in the European Union. *Journal of European Public Policy* 19(4): 454–471.

Boussena, S., Locatelli, C., 2013. Energy institutional and organizational changes in EU and Russia: Revisiting gas relations. *Energy Policy* 55: 180–189.

Cavaliere, A., 2007. *The Liberalization of Natural Gas Markets: Regulatory Reform and Competition Failures in Italy.* Oxford: Oxford Institute of Energy Studies.

CIEP, 2011. *CIEP Vision on the Gas Target Model.* The Hague: Clingendael International Energy Programme. See www.clingendaelenergy.com/inc/upload/files/Gas_Target_Model.pdf.

Correljé, A., De Jong, D., De Jong, J., 2009. *Crossing Borders in European Gas Networks: The Missing Links*. The Hague: Clingendael International Energy Programme.

Creti, A., Villeneuve, B., 2005. Longterm contracts and take-or-pay clauses in natural gas markets. *Energy Studies Review* 13(1): article 1.

De Hautecloque, A., Glachant, J.-M., 2009. Long-term energy supply contracts in European competition policy: fuzzy not crazy. *Energy Policy* 37(12): 5399–5407.

Dorigoni, S., Portatadino, S., 2008. LNG development across Europe: infrastructural and regulatory analysis. *Energy Policy* 36(9): 3366–3373.

Dorigoni, S., Graziano, C., Pontoni, F., 2010. Can LNG increase competitiveness in the natural gas market? *Energy Policy* 38(12): 7653–7664.

European Commission, 2007. *Communication: Inquiry Pursuant to Article 17 of Regulation (EC) No. 1/2003 into the European Gas and Electricity Sectors (Final Report)*. COM(2006) 851 final. Brussels: European Commission.

European Commission, 2010. *Communication: Energy Infrastructure Priorities for 2020 and Beyond: A Blueprint for an Integrated European Energy Network*. COM(2010) 677 final. Brussels: European Commission.

European Commission, 2011. *Proposal for a Regulation on Guidelines for Trans-European Energy Infrastructure and Repealing Decision Number 1364/2006/EC*. COM(2011) 658 final. Brussels: European Commission. See http://eur-lex.europa.eu/LexUriServ/LexUriServ.do?uri=COM:2011:0658:FIN: EN:PDF.

European Commission, 2012. *Communication: Making the Internal Energy Market Work*. COM(2012) 663 final. Brussels: European Commission.

European Commission, 2014a. *Communication on the Short Term Resilience of the European Gas System*. COM(2014) 654 final. Brussels: European Commission.

European Commission, 2014b. *Communication: European Energy Security Strategy*. COM(2014) 330 final. Brussels: European Commission.

Everis and Mercados EMI, 2010. *From Regional Markets to a Single European Market*. Brussels: European Commission. See http://ec.europa.eu/energy/sites/ener/files/documents/2010_gas_electricity_markets.pdf.

Finon, D., Roques, F., 2008. Financing arrangements and industrial organization for new nuclear build in electricity markets. *Competition and Regulation in Network Industries* 9(3): 247–281.

GIIGNL, 2012. *The LNG Industry in 2012*. Paris: Groupe International des Importateurs de Gaz Naturel Liquéfié.

Glachant, J.-M., 2011. *A Vision For the EU Gas Target Model: The MECO-S Model*. Working paper RSCAS 2011/38. San Domenico di Fiesole: Robert Schuman Center for Advanced Studies, European University Institute.

Glachant, J.-M., Hallack, M., Vazquez, M., 2013. *Building Competitive Gas Markets in the EU: Regulation, Supply, and Demand*. Loyola de Palacio Series on European Energy Policy. Cheltenham: Edward Elgar Publishing.

Goldthau, A., Boersma, T., 2014. The 2014 Ukraine-Russia crisis: implications for energy markets and scholarship. *Energy Research and Social Science* 3: 13–15.

Harmsen, R., Jepma, C. J., 2011. North west European gas market: integrated already. *European Energy Review* January 27.

Hartley, P., 2013. *The Future of Long-Term LNG Contracts*. Cambridge, MA:

Belfer Center, Harvard University/Houston, TX: Baker Institute for Energy Studies, Rice University.

Heather, P., 2012. *Continental European Gas Hubs: Are They Fit for Purpose?* Oxford: Oxford Institute for Energy Studies.

Helm, D., 2014. The European framework for energy and climate policies. *Energy Policy* 64: 29–35.

Henderson, J., 2012. *The Potential Impacts of North American LNG Exports.* Oxford: Oxford Institute for Energy Studies.

Johnson, C., Boersma, T., 2013. Energy (in)security in Poland, the case of shale gas. *Energy Policy* 53: 389–399.

Kalicki, J. H., Goldwyn, D. L., (eds) 2013. *Energy and Security: Strategies for a World in Transition*, second edition. Washington, DC: Woodrow Wilson Centre Press.

Katz, S., Jepma, C. J., 2012. Ranking European gas markets. *European Energy Review* October 23.

Kumar, S., Kwon, H.-T., Choi, K.-H., Cho, J.H., Lim, W., Moon, I., 2011. Current status and future projections of LNG demand and supplies: a global perspective. *Energy Policy* 39(7): 4097–4104.

Lochner, S., Bothe, D., 2009. The development of natural gas supply costs of Europe, the United States and Japan in a globalizing gas market: model-based analysis until 2030. *Energy Policy* 37(4): 1518–1528.

Makholm, J. D., 2012. *The Political Economy of Pipelines: A Century of Comparative Institutional Development.* Chicago, IL: University of Chicago Press.

Neuhoff, K., Von Hirschhausen, C., 2005. *Long-term vs. Short-term Contracts: A European Perspective on Natural Gas.* Working paper (preliminary research findings), CWPE 0539 and EPRG 05. See www.repository.cam.ac.uk/bitstream/handle/1810/131595/eprg0505.pdf?sequence=1.

Nowak, B., 2010. Equal access to the energy infrastructure as a precondition to promote competition in the energy market: the case of the European Union. *Energy Policy* 38(7): 3691–3700.

Paltsev, S., Jacoby, H. D., Reilly, J. M., Ejaz, Q. J., Morris, J., O'Sullivan, F., Rausch, S., Winchester, N., Kragha, O., 2011. The future of US natural gas production, use, and trade. *Energy Policy* 39(9): 5309–5321.

Panke, D., 2007. The European Court of Justice as an agent of Europeanization? Restoring compliance with EU law. *Journal of European Public Policy* 14(6): 847–866.

Paraja, M. A. R., 2010. New infrastructures in the Spanish Gas Market. CNE presentation, FSR and BnetzA Forum on Legal Issues of Energy Regulation, Florence School of Regulation, October 15.

Pearson, I., Zeniewski, P., Gracceva, F., Zastera, P., McGlade, C., Sorrell, S., Speirs, J., Thonhauser, G., Alecu, C., Eriksson, A., Toft, P., Schuetz, M., 2012. *Unconventional Gas: Potential Energy Market Impacts in the European Union.* JRC Scientific and Policy Reports. Brussels: European Commission.

Renou-Maissant, P., 2012. Toward the integration of European natural gas markets: a time-varying approach. *Energy Policy* 51: 779–790.

Smith, W. J., 2013. Projecting EU demand for natural gas to 2030: a meta-analysis. *Energy Policy* 58: 163–176.

Spanjer, A. R., 2009. Regulatory intervention on the dynamic European gas market:

neoclassical economics or transaction cost economics? *Energy Policy* 37(8): 3250–3258.

Stephenson, E., Doukas, A., Shaw, K., 2012. Greenwashing gas: might a "transition fuel" label legitimize carbon-intensive natural gas development? *Energy Policy* 46: 452–459.

Stern, J. P., 2007. Is there a rationale for the continuing link to oil product prices in continental european long-term gas contracts? *International Journal of Energy Sector Management* 1(3): 221–239.

Stern, J., 2009. *Continental European Long-Term Gas Contracts: Is a Transition Away from Oil Product-Linked Pricing Inevitable and Imminent?* Oxford: Oxford Institute for Energy Studies.

Stern, J., 2014. International gas pricing in Europe and Asia: a crisis of fundamentals. *Energy Policy* 64: 43–48.

Stern, J., Rogers, H., 2011. *The Transition to Hub-Based Gas Pricing in Continental Europe.* Oxford: Oxford Institute for Energy Studies.

Steunenberg, B., Toshkov, D., 2009. Comparing transposition in the 27 member states of the EU: the impact of discretion and legal fit. *Journal of European Public Policy* 16(7): 951–970.

Vazquez, M., Hallack, M., Glachant, J.-M., 2012. *Building Gas Markets: US versus EU, Market versus Market Model.* Working paper RSCAS 2012/10. San Domenico di Fiesole: Robert Schuman Center for Advanced Studies, European University Institute.

Von Hirschhausen, C., 2008. Infrastructure, regulation, investment and security of supply: a case study of the restructured US natural gas market. *Utilities Policy* 16(1): 1–10.

Von Hirschhausen, C., Neumann, A., 2008. Long-term contracts and asset specificity revisited: an empirical analysis of producer–importer relations in the natural gas industry. *Review of Industrial Organization* 32: 131–143.

Von Hirschhausen, C., Neumann, A., Ruester, S., Auerswald, D., 2008. *Advice on the Opportunity to Set Up an Action Plan for the Promotion of LNG Chain Investments: Economic, Market, and Financial Point of View.* Final report. Brussels: DG-TREN, European Commission.

Zhelyazkova, A., 2013. Complying with EU directives' requirements: the link between EU decision-making and the correct transposition of EU provisions. *Journal of European Public Policy* 20(5): 702–721.

Part IV
Conclusions

8 Conclusions and recommendations

A long and bumpy road to European Union energy security

This chapter contains the conclusions from the previous chapters as well as recommendations in terms of policy making and future research. It has three parts, which together provide answers to the research questions raised at the beginning of this book. The first part deals with the main theme of this research (i.e. EU energy security considerations arising from the internal gas system). The second part examines European energy policy related issues and gives recommendations for future policy making and further research. The chapter ends with some brief reflections.

European Union energy security

This book assessed whether EU energy supply is at risk because too many decisions are taken at a suboptimal level of policy making. In short, the evidence suggests that this is the case for parts of the EU. This section presents the evidence that supporting this conclusion.

The analysis of investments in gas infrastructure (Chapter 5) demonstrates that the EU and US regulatory regimes have a fundamentally different approach towards energy infrastructure investments (see also Vazquez et al. 2012). In the US regulation is used as an investment vehicle (Joskow 2005) whereas in the EU regulation focuses on efficiency, rather than security of supply or sustainability. As a result, investors in gas transmission infrastructure in the US are allowed to accrue substantially higher rates of return than their European counterparts. In addition, regulatory periods in Europe are, being limited to three or four years, comparatively short and because of that returns on investment can vary substantially. The EC estimates that the necessary investments in gas infrastructure (of approximately €70 billion in the period up to 2020) are not made under business as usual conditions (e.g. European Commission 2010, 2012). More research is needed whether in the US consumers, as a result of this generous regulatory regime, collectively pay too much for their gas transportation. Furthermore, in the US there are millions of citizens that are not connected to gas distribution grids. Instead, these people use alternative fuel sources such as propane or wood. Therefore, it seems fair to presume that in the US non-economical pipelines are not being built as a result of this market oriented approach. Contrary to the US, the EU does not have an institution that orchestrates interstate pipelines and cross-border invest-ments, which therefore do not materialize in all parts of the EU. Academic evidence does not provide a clear verdict in favor or against privatization of gas networks. It is worth noting that throughout Europe progress has been made in the last half decade to further integrate markets, most importantly because EU institutions are increasingly acknowledging the importance of integrating European gas networks, as a means to increase resilience to market abuse and arbitrary pricing by dominant suppliers.

Chapter 5 also provides an illustration of (the struggles of) the European integration process. Since regulation of the transport of natural gas has predominantly been a responsibility of the member states, it is hardly surprising that as a result Europe has a patchwork of different regulatory regimes (Larsen *et al.* 2006). This study gives several illustrations of the inefficiencies this has created. It is important to note that the heterogeneity of regulatory regimes throughout Europe in itself does not seem to be a major impediment to investments in infrastructure, though that changes when cross-border projects such as interconnectors are involved (e.g. Ruester *et al.* 2012). The lack of international cooperation was mentioned explicitly by the EC in light of the Ukraine crisis and the so-called stress-tests (European Commission 2014a). In order to streamline national regulations in 2010 the Agency for the Cooperation of Energy Regulators (ACER) was established. Yet assessments of this organization have been critical, with studies concluding that it lacks decision-making powers, the right to set its agenda, that it has a mediocre budget, and that it is a "second best" option (Coen and Thatcher 2008; Correljé *et al.* 2009; Thatcher 2011). It may be that some neofunctionalist scholars have been too opti-mistic in their assessment that European integration will eventually proceed in any case, even if that sometimes means taking a step back (Corbey 1995). By now all stakeholders involved are aware of this process and deliberately may have stopped cooperation. It is difficult to ascertain what exactly the reasons are that influence this process. Possibly the more modest role of ACER in comparison to its peers in the member states in the first phase after its foundation is part of conventional decision-making procedures in Europe. It could be that a large mandate at first would generate too much resistance, since ultimately enlarging ACER's mandate would also mean reducing the influence of national regulatory authorities. It is reasonable to assume that academic and political verdicts of ACER would benefit from considering this logic. In general European integration is a lengthy process, without a preconceived objective. The political–economic–technological dynamics of energy-related issues in Europe should be assessed in this specific context. In other words, it is worth appreciating the unique setting in which European integration takes place. As such, the establishment of ACER as a European organization can also be seen as a substantial milestone, for it marks the acknowledgement that designing European energy regulation strictly at the member state level leads to undesirable inefficiencies. On the other hand, it is likely that the transfer of decision-making powers from the member state level to ACER will be a lengthy process, if it happens at all. The period in between may be one of relative uncertainty and unpredictability, conditions that generally do not improve the investment climate. This fact has led some to conclude that completing Europe's internal market may take several decades (Makholm 2012). CIEP (2011) has argued that the EU cannot afford to take such a long time, when global competition for energy resources is growing.

In addition, in Chapter 5 questions are raised about the political and legal accountability of ACER (e.g. Lavrijssen-Heijmans and Hancher, in Arts *et al.* 2008). Similar concerns have been expressed with regard to national independent regulatory authorities. The oft-heard argument is that creating these independent agencies shifts complex technical issues out of the political arena (Elgie 2006). The majority of studies, however, question the independent positions of these agencies in terms of democratic legitimacy and a lack of accountability (Larsen *et al.* 2006; Christensen and Laegreid 2007; Maggetti 2009). Szydlo (2012) even argues that economic and social goals of regulatory authorities often collide, and that depriving national parliaments of legislative influence is in conflict with constitutional principles, such as the domain of the law. It appears that more research into the independent position of regulatory authorities is desirable.

The case study on shale gas extraction (Chapter 6) demonstrates the transformation that the US is undergoing in terms of domestic natural gas production. Clear effects have been identified on the ground in for instance rural areas of Pennsylvania, where once forgotten towns are booming again, new roads are being constructed to accommodate intensive truck usage, hotels built to house the workforce, and local jobs created. It is worth noting that the available empirical evidence suggests that earlier predictions about job creation have been too optimistic (Weber 2012). Nevertheless, gas production has increased very significantly and the US is expected to become a potential net exporter of natural gas in 2015, whereas until roughly 2008 most observers expected it to become a major importer of LNG. On the other hand the analysis also lays out several substantial environmental concerns that have been linked to shale gas extraction, most notably drinking water contamination, waste water treatment, air pollution, and induced seismic activity. Effective regulation of these environmental concerns at both the federal and the state level have so far generated mixed results. It is possible that the lack of academic consensus on the environmental risks and their links with shale gas extraction have further delayed effective environmental regulation. Finally it is worth noting that in the US the extraction of shale gas is not embraced in full. The legal bans on hydraulic fracturing (the currently preferred technology to extract the natural gas from shale rock layers) in states like Vermont and also New

York, and also in local communities in states as diverse as Ohio, Colorado and Texas, are good examples of the skepticism amongst a part of the American populous. Overall, however, broad public support has been identified as one of the enablers of what has been labeled the shale gas revolution (Boersma and Johnson 2013).

Contrary to the US, in Europe shale gas extraction is in an embryonic phase. To date, not a single molecule of shale gas has been produced. Similar to the US, in Europe opinions vary widely, from legal bans in France and Bulgaria to excitement in Poland and active government support in the United Kingdom. However, as the analysis shows, one of the member states that most enthusiastically wants to start extracting shale gas from its soils, Poland, has substantial internal hurdles to overcome. Even if geologic conditions turn out to be favorable, at the moment the lack of domestic market development, insufficient interconnection facilities and available infrastructure and the meager implementation of existing European legislation would hinder large-scale exploitation of this natural resource (Johnson and Boersma 2013). At this point in time the initial optimism about shale gas in Poland has reduced significantly. Indeed, by December 2014 five private sector companies left the country to start exploration elsewhere, because of disappointing test drillings, but also owing to disappointment with ongoing regulatory uncertainty.

Similar to the case study in Chapter 5, this case study also outlines the dynamics (and complications) of European integration. Comparable to the US, in Europe resource extraction is the exquisite domain of the member states. Unlike in the US, in Europe regulation of environmental concerns, for instance air quality, water quality, and noise pollution, are dealt with at the supranational level. Hence, while European institutions have declared repeatedly to be "neutral" with regard to shale gas extraction (European Commission 2012), member states such as Poland have been awaiting assessments from Brussels regarding the existing frameworks to safeguard environmental concerns. To date it remains unclear whether the existing frameworks are deemed sufficient to address the environmental concerns, in case of large scale extraction of shale gas. The non-binding recommendation that was published by the EC in 2014 only adds to that uncertainty, as it effectively postpones a decision about possible binding regulation on shale gas until 2016 (see Boersma and Khodabakhsh 2014). It is difficult to assess why European member states have accepted supranational interference with environmental concerns, while at the same time resource extraction is an exclusive domain of the member states. It may be that advocates of environmental protection have acknowledged that their voices are best represented at the supranational level, since many of these concerns (e.g. air pollution or water-related issues) do not have clear boundaries and may therefore be addressed more effectively in a broader policy initiative. Advocates of energy industries to the contrary have a strong foothold in their respective member states, as do infrastructural companies and regulatory authorities. However, this study demonstrated that the member state level may not always be the most efficient level to deal with energy policy related issues. Despite this functional logic, transfer of power from the member state level is a lengthy process. There can be multiple reasons for this, such as the unwillingness of national interest groups to give up power, or the national focus of most lobby groups in the member states. More research is needed to ascertain the exact nature of these dynamics and their consequences.

Given the novelty of shale (and more broadly unconventional) gas extraction in the world, many issues remain unexplored. These issues comprise, for instance, the long-term consequences for markets in the US and elsewhere in the world, uncertainties regarding environmental risks and the effective regulation thereof, technological progress and innovation, and also the geopolitical consequences of this major shift in the energy landscape.

The study of Europe's market structure (Chapter 7) examines several components of Europe's gas system. The analysis shows the need for substantial investments in gas infrastructure (€70 billion) throughout the continent in the period up to 2020. Contrary to the US where ACER plays this role, in Europe there is no institute to orchestrate that the necessary investments, in particular in cross-border infrastructure, are being made. This would not be problematic, if member states more efficiently coordinated these investments, yet this is not always the case. In fact, there is substantial evidence that gas infrastructure investments with a strong cross-border component are not being made (Ruester *et al*. 2012; European Commission 2014a). As a result, Europe's gas system is developing at different paces, with northwestern Europe reasonably well integrated, while the rest of Europe falls behind (Abada and Massol 2011; Renou-Maissant 2012; Asche *et al*. 2013). Several studies have linked this asynchronous development to a lack of coordination/integration in the EU energy system (Spanjer 2009; Correljé *et al*. 2009; Vazquez *et al*. 2012). The analysis also suggested that the expected decision regarding critical infrastructure (providing European institutions for the first time with a structural budget for energy infrastructure) will not substantially alter the status quo. This fact arises because the final budget for the Connecting Europe Facility is €5.85 billion for energy infrastructure (both natural gas and electricity) for the period up to 2020 and therefore only comprises a fraction of the estimated costs. Moreover, the budget allocation for particular projects is going to be political, and therefore it is an easy subject for criticism. Surely recently adopted legislation on energy infrastructure and its financing under the Connecting Europe Facility are a push in the right direction, but there is no reason to believe that they will be a panacea. With that in mind, the development of plans for an Energy Union under EC President Juncker in 2015 and beyond are important to watch, as energy security has become one of the center pieces of his presidency.

Second, the study provides an overview of pending infringement proceedings, as of late 2012. The overview shows that by that time fourteen member states had not implemented existing legislation, which should have been implemented in the spring of 2011. It is also worth noting that this may not be the conclusive list of noncompliant EU member states. Academic literature shows that cases are solved frequently soon after the EC starts an infringement procedure, but it unclear whether that is also the case when more than half of the member states is noncompliant. Though anecdotal evidence suggests that in late 2014 the number of noncompliant member states had diminished to six or seven, arguably that is not the end of the problem. More analysis would be useful here to prove or disprove the above points.

Third, Chapter 7 analyzes the development of gas market trade in the EU. Again substantial differences within the EU have been identified. In northwestern Europe market trade is reasonably well developed (Harmsen and Jepma 2011; Heather 2012; Stern 2014), yet in other parts of Europe oil-indexed long-term contracts prevail (Abada and Massol 2011). The analysis of price developments at four important global benchmarks (*inter alia* British NBP and the German border) seems to confirm that the reported oversupply of natural gas in 2008 made the difference between spot-market prices and long-term contracts so large, that renegotiations of existing contracts took place (Stern 2009). Several studies note the advantages of long-term contracts for investments and stability, at a hidden cost for society (De Hautecloque and Glachant 2009). The EC has expressed its support for hub-based trading (European Commission 2012), yet seems to have the intention to let spot-market prices and long-term contracts co-exist. Hence it is difficult to assess what the future of long-term contracts in Europe will be. One option is that these contracts in the future will increasingly be indexed on spot-market natural gas prices. It is also worth reiterating that while declining spot-market prices have had a significant effect on markets in northwestern Europe, the impacts on markets in central and eastern Europe have been more modest, due to the lack of liquid markets, trade, and physical infrastructure bottlenecks. Here, long-term contracts continue to prevail, even though most buyers have received temporary discounts under pressure of declining global gas prices. It seems that only the Czech Republic, due to its improved connection to the German market, has benefitted in terms of declining wholesale prices (Noël, in Kalicki and Goldwyn 2013).

The final part of Chapter 7 analyzes the role of LNG in Europe's gas system. The study shows that substantial investments in LNG regasification terminals have been made throughout Europe. As a result in August 2014 Europe had 22 operational LNG terminals, with four under construction, compared to twelve operational terminals in 2006. However, scholars do not agree on the future role of LNG in Europe, with some expecting an increase in LNG's share in Europe's energy mix (Dorigoni *et al.* 2010; Kumar *et al.* 2011; Boersma *et al.* 2014), while others predict a decline (Lochner and Bothe 2009). An analysis of industry data suggests that the center of gravity for LNG demand has been shifting further to Asia (due to increased LNG demand in Japan after Fukushima, and also upcoming LNG markets such as China and India), while Europe's share fell from 27 percent in 2011 to 20 percent in 2012 (GIIGNL 2012). More research would be useful here, since the future of LNG in Europe depends on a complex set of factors (e.g. future pricing mechanisms, carbon policies, and the size of streams of natural gas that may come online in Australia, Russia, east Africa, and North America). Finally, an overview of infrastructural bottle-necks in central and eastern Europe suggests that it is currently uncertain whether this region could benefit from more LNG supplies in Europe, since natural gas cannot always flow to and through central and eastern Europe. Finally, it is too early to tell what the effects will be of LNG terminals that are expected to start operations from 2015 onward in both Poland and Lithuania. There is reasonable certainty that the economics of LNG in this part of the EU gas system have to be questioned, in both cases because the contracted LNG (from Qatar and Norway respectively) is more expensive than currently contracted Russian natural gas. Anecdotal evidence suggests that Polish policy makers have decided that LNG prices will be regulated, because otherwise there would be no market for them. Indirectly, the Polish tax payer is thus paying a premium for alternative gas supplies. In the case of Lithuania, allegedly the law has been amended so that buyers of natural gas in the country are forced to purchase LNG before purchasing pipeline gas (from Russia). More empirical work is required to verify these observa-tions, and to assess their long-term effects on the EU gas market functioning more broadly.

This third case study also shows the complex process of European inte-gration, as it touches upon the often postponed decision about the European proposal for a regulation on energy infrastructure. Despite the fact that this legislation was first proposed in the fall of 2011, it took until mid-2013 to be adopted, even though the European Council explicitly requested this legislation from the EC. Only in October 2014 the first three projects were published that had been selected for co-investment. However, since its first publication, there have been questions from member state representatives about the role of European institutions with regard to the financing of energy infrastructure. Since northwestern Europe has reason-ably well developed gas infrastructure facilities, stakeholders from these member states often oppose European interference with energy infrastruc-ture financing. Other parts of Europe may benefit from European funds to develop their domestic gas market, and are therefore in favor. As outlined, in the illustration of the European Energy Program for Recovery in Chapter 3, in the rare case that European institutions have a mandate regarding investments in energy infrastructure, all member states want a piece of the pie. Yet with the asynchronous development of parts of the European gas system, this mechanism does not necessarily contribute to further integra-tion of the EU energy system as a whole. Ideally the money would be spent where it is most urgently needed, but in reality allocation may well turn into a heavily politically inspired, and therefore contentious, process.

A review of energy security literature demonstrates that scholars overwhelmingly focus on the availability of supplies and the reliability of supply routes to Europe. The debate also shows that other vital components of the energy system, notably available infrastructure (interconnections, reserve flows) and regulations are not as embedded in the energy security discourse. This leads to several questions about Europe's attempts to address energy security. First, despite the academic focus on available supplies and the substantial amount of political discussion on this topic, Europe lacks a coordinated approach. Instead, its relations with some of its key external suppliers are troublesome. It raises the question how valuable the concept of security and its debate are to deal with energy related issues. Or does the security debate mainly function as a welcome distraction of the fact that internal cooperation is just not progressing swiftly? Second, while other vital components of the energy system have not been embedded in the security discourse, this study highlighted numerous examples of how difficult European integration is proceeding in these issue areas as well. To give an example, despite the urgent need to invest billions of euros in additional infrastructural projects, European member states continue to struggle with questions about who has the mandate to carry out such operations and also who is going to pay for them. In sum, European energy security concerns have both an internal and an external dimension. This study has focused mostly on the internal dimension and demonstrates that the EU is currently not agile enough to improve the internal inefficiencies that hinder natural gas to flow throughout the continent. In Chapter 4 this study also examined the case of Regulation 994/2010, which aims to increase the EU mandate to address energy security from Brussels. Even though the regulation is only 5 years old and there has not been a significant supply disruption since, the Ukraine crisis in 2014 provided a good test to see whether this regulation has contributed to enhance energy security in the EU. As discussed, its contribution can only be called modest. On the upside, the regulation provided European policy makers with more data and in general created a more transparent gas system. At the same time, it is striking to see how upon evaluation in 2014 several member states continue to fail to implement basic European legislation, most notably the countries that are most outspoken when it comes to energy security and an existential threat called Russian natural gas supplies. If in particular those member states that claim to see this as a major threat make comparatively modest efforts to address the problem, one must question whether that problem is as grave as portrayed. The literature suggests that with regard to the external dimension of the discussion on security of supply, the EU equally struggles with its policy targets, as for instance indicated by the ongoing debates about the Third Energy Package or the role of long-term contracts, which has already adversely affected the EU-Russian relationship and may deteriorate it further over the course of 2015.

European Union energy policy

The analysis in Chapter 3 contains the most prominent European legislative documents related to the European gas system that have been drafted in Brussels to date. Perchance in the spirit of the 1990s, policy makers' efforts have focused initially on market liberalization and development. Subsequently the attention shifted to other crucial parts of the gas system (i.e. infrastructure and regulation). The analysis shows that decision-making powers regarding infrastructure and regulation predominantly reside at the member state level.

Chapter 3 suggests that in policy making a certain mismatch can be identified, which seems almost characteristic for the development of the European gas system. Some authors have pointed to the Lisbon Treaty's provisions on shared competence of energy policy between member states and European institutions as adopted in 2010 (Trombetta 2012). Yet it remains to be seen what this means in reality, since that same Lisbon Treaty makes clear: interference with national sovereignty would automatically block initiatives of supranational nature.

This analysis puts forth several suggestions on how completion of the internal gas system could be accelerated. First, implementation structures for existing legislation are not always functioning. Although this may be expected from, for example, member states in eastern Europe that only joined the EU in 2004, as of late 2012 the EC reported fourteen pending infringement procedures throughout the continent. Therefore an existing or new authority to orchestrate implementation, with punitive powers, could be helpful. Furthermore completing the internal energy system would require that regulations across the EU are better streamlined where necessary, since the current patchwork is non-transparent and not efficient. Also new incentives are needed to generate attract investments in gas infrastructure facilities, albeit transmission lines, storage facilities or enhancement of increased interconnection and reverse flow technology, with a focus on projects that have a strong cross-border component. An institution to orchestrate these issues could very well be ACER, though its mandate would need to be expanded. Alternatively better coordination between member states could be strived for, though the institutional history that was touched upon in this analysis does not suggest that would be an effective route, as was recently confirmed in several analyses of the EC in light of the Ukraine crisis (European Commission 2014a, 2014b). More research is needed into how to attract larger financing for infrastructure and mechanisms to invest the necessary billions of euros in the decade ahead. It may be that increasingly public financial means are needed to develop parts of the EU system, in particular in eastern and southeastern Europe, for these markets may not be mature enough to attract sufficient private capital. A more prominent role for the European Investment Bank could be considered. At the same time, other measures have to be considered in order to avoid free-rider behavior (e.g. extending regulatory periods, temporary increases in rates of return to stimulate investments, and creating stimuli focused particularly on projects with a strong cross-border component). The previous section discussed how the US gas system may contain valuable lessons in this respect.

Though this book has analyzed the internal EU gas system, a study of EU energy security inevitably also touches upon external suppliers. This is because Europe has always been highly dependent on external suppliers for its natural gas. In that sense, the continuous difficult political relationship between the EU and Russia does not match one of the basic functionalist principles: form follows function. A necessary related question is: function for whom? Even with clear interests of stable relations on both sides of the equation, in terms of security of supplies and security of demand, EU/Russian relations remain burdensome, with some studies suggesting that the two are further drifting apart (Boussena and Locatelli 2013) even before the Ukraine crisis started in 2014, which may have deteriorated EU/Russian relations for a long time to come. This is remarkable, given the obvious interdependence between the two. As described, domestic production of natural gas in Europe, including potential natural gas from shale rock layers, is not expected to make up for even half of its consumption (Pearson *et al.* 2012). Despite the increase of natural gas that is produced globally and the increasing globalization of gas markets, it may be expected that prices dictate that the cheapest gas finds its way to Europe. Generally, that would be conventional natural gas transported through pipelines from its closest suppliers (Lochner and Bothe 2009; Paltsev *et al.* 2011; Boersma *et al.* 2014). Considering the limited options of Norwegian supplies to be ramped up, and also ongoing political turmoil in and uncertainty about future gas production in Algeria, large amounts of Russian gas in Europe's future energy mix seem inevitable. Substantial challenges thus have to be overcome to improve European–Russian relations. Continued quarrels over the Europe's Third Package, ongoing struggles of Gazprom to adjust itself to the new market realities of lower gas tariffs, more spot market trade, and renegotiation of long-term contracts, and concerns (valid or not) about the lack of investments in so-called "green fields" in Russia are witness to that. Furthermore, in light of the Ukraine crisis increased collaboration in gas trade between Russia and China has to be taken into consideration, although the details of the agreements made have to be ironed out and stable gas trade between Russia and China is expected to take several years.

Given this explicit external dimension of future European energy security, however, further strengthened by large transit pipelines, investing in stable relations seems potentially beneficial to both parties. The withdrawal of Russia from the Energy Charter Treaty in 2009 and the ongoing legal quarrels over the Third Package mark exactly the opposite direction. The external dimension of European energy security confirms that the EU is in fact a suboptimal level to govern the European gas market. From a functional perspective, large and nearby suppliers of natural gas (i.e. Russia, Norway, and Algeria) should be an integral part of the relevant governance structures. Since within Europe there is no coherent external energy policy, as a consequence relations between these suppliers and Europe have evolved in different ways. This seems to have led to little debate when dependence on Norway and also Algeria is concerned. In the case of Russia, to the contrary, since 2006 the political debate about the reliability of the country as Europe's largest natural gas supplier has bloomed. Arguably the two major supply disruptions in 2006 and 2009 have caused uncomfortable situations for substantial amounts of Europeans citizens. Yet negative sentiments surrounding Russia have also functioned as a distraction from problems related to the EU internal gas market. As this study has shown, the proclaimed security threats often have to be taken with a grain of salt. Also, the review of academic literature on energy security suggests that a substantial amount of contributions have faced this pitfall by focusing exclusively on the external dimension of energy security, whereas some of its acute challenges are *within* the European gas market.

The above issues are vital to consider when designing the future EU gas system. The results of this study have suggested that it may indeed be difficult to "shoehorn" this system into a particular desired market model, so choosing its own path seems to be the only way forward for Europe. More empirical work is required to establish what exactly that path would comprise, but it is certain that substantial import dependency with regard to natural gas is one key element that has to be taken into consideration.

Reflection

Several aspects of this manuscript deserve a short reflection. First and foremost the concept of energy security is subject to continuous debate. The discussion sometimes almost takes religious forms, in which facts are not always adequately presented. The securitization of energy resources, or politicization in many instances, has not been helpful. Though some commentators have hinted at the differences between some of the EU member states (where eastern European states are generally reported to have a higher degree of securitization of energy resources than states in northwestern Europe) here too it is worth noting a certain degree of rhetoric. For example, it is unclear why, in the case of Poland, when Russia is portrayed as an existential threat, investments have not been made earlier to prepare Poland to receive natural gas from different suppliers. Poland could have invested in a pipeline with Denmark to receive Norwegian gas, or in interconnection and reverse flow facilities with Germany to improve its options to receive natural gas from northwestern Europe, and more importantly, it could have started investing in 2006, when this issue became apparent. The fact that is has chosen to only modestly invest in alternatives, and that it chose the least efficient option to diversify its natural gas supplies (i.e. an LNG terminal requiring price regulation because otherwise the resource would not be competitive), demonstrates the level of rhetoric in energy security debates in Europe. In addition it confirms the national approach that member states generally take when designing measures to address energy security. It also demonstrates that though the tone of these debates may sometimes suggest a high level of securitization, in fact political actions or willingness to act tell a different story. More research is needed to unravel the reasons behind this difference between the expressed concerns and factual behavior.

Another problematic feature of the concept of energy security is the myriad of different interpretations. As this study has shown, in general most attention is paid to available supplies and reliability of suppliers, and less is written about other crucial features of the gas system (i.e. infrastructure and regulatory authorities). Some authors have called for a common standard of energy security, but this seems impossible given the political dimension to energy resources, differences in interests per actor as well as changes over time. The good news, however, is that while over time the interest of scholars in energy security evaporates, it returns after the next supply disruption or crisis. More research is needed to address the question why so many scholars discuss energy resources in terms of a security matter. This analysis suggests that this framing is not necessarily helpful in addressing energy-resource-related questions.

The theoretical framework requires several reflections. Europe's dealing with energy security corresponds only to a limited extent with insights from (neo)functionalist theory. Though the establishment of the internal market may seem in line with functionalist thinking, in practice this is questionable. This study has shown that borders are still relevant, also within the EU, which for instance becomes apparent through the lack of physical interconnection capacity. It seems that even when it is specified for whom something can be functional and to what purpose (e.g. further market integration to create alternatives in case of external supply disruptions) market integration may be hindered. Thus, when addressing the basic notion whether form follows function, the results of this study suggest that this is currently not always the case: the EU can be explicit about form and the question who is in and who is out, and the difficult relations with Russia are a prime example. The relations with another major external supplier, however, Norway, suggest a different narrative. Here "function" does prevail, and form (being part of the EU or not) seems to be less relevant. Thus, while beyond the scope of this study, more comparative research into the dynamics of relations between the EU and its three major suppliers of natural gas, Russia, Norway, and Algeria, would be useful and possibly provide an interesting exploration of (the limits to) neofunctionalist thinking. Insights from new institutional economics have been helpful in this study, for the research focused predominantly on the EU internal gas market. Therefore this study partly sought to address institutions and context, elements that are crucial to economic activity and that these theoretical contributions focus on. They also contain an imminent component, for changing the formal rules of the game (legislation, regulations) is expected to take decades (e.g. Williamson 2000). It is, however, unclear whether this is always the case. If so, what would that mean in terms of European energy market development and related security questions? It is worth reiterating that this stream of research is comparatively new and that many issues related to institutions yet have not been discovered. Finally multilevel governance as a framework of analysis has been valuable since it acknowledges the role of actors other than states in governance structures. The distinction between public and private actors was also relevant and helpful in this study. Finally the framework proved applicable to the case of the US, which was used as a benchmark. What this study did not do, is test MLG as a theory of European integration, because that debate was beyond the scope of this research project.

The case studies in this book merit several observations. First, the case study on investments in gas infrastructure focused exclusively on US interstate gas pipelines. Therefore more work is needed on for instance intrastate pipelines, for this may bring about additional and perchance different results. Moreover one could argue that intrastate pipelines in the US in a geographical sense show remarkable resemblance with most European gas pipelines, if only since these have predominantly been organized at the member state level. Although existing academic contributions do not suggest that this is in fact a flaw in this research design, more empirical work would be useful here. In that same study more work could be done to examine the investment rationale in other EU member states than the two under study here. Although academic studies confirm that Great Britain and also the Netherlands can be seen as frontrunners in the EU with regard to energy regulation issues, more in-depth work on practices in other member states may bring about valuable lessons on for instance incentives to increase investments in gas infrastructure. One caveat that applies to the entire case study on shale gas extraction is that this field is new and unexplored. Therefore next to offering opportunities to carry out new research, many issues remain undiscovered for now. To give an example, despite the literally thousands of wells that have been drilled every year so far in the Marcellus shale, the largest and most promising shale rock layer in the US, up to now it is highly uncertain how much natural gas can eventually be economically recovered from it. That has to do with the size of the rock layer, covering large parts of Ohio, Pennsylvania, West Virginia, and New York, but also with the fact that in one of those states (New York) hydraulic fracturing has not been permitted. Therefore, and because of possible future technological developments, relatively little can be said about recoverable reserves in the Marcellus shale (though it is certain that these will be significant). Logically, in comparison to the US there are no data available in the EU about recoverable reserves of natural gas from shale rock layers, making future predictions about its role on this continent highly uncertain. Over time, as more data are expected to emerge, many questions remain regarding the consequences of shale gas for markets, its currently fiercely debated environmental consequences, effects on other natural resources such as water, and also its geopolitical consequences. As for the last case study on market structure, it reiterates how much work remains to be done to construct the EU energy system (while at the same time highlighting the significant progress that has been made), while it also laid out some of the political challenges that inevitably follow from the geographical location of the continent. Tucked in between large external suppliers of natural gas, it leaves a tremendous responsibility for academics, policy makers and business representatives alike to shape Europe's energy system and construct external relations in a fashion that facilitates secure and affordable supplies on the way to a post-carbon era.

References

Abada, I., Massol, O., 2011. Security of supply and retail competition in the European gas market: some model-based insights. *Energy Policy* 39(7): 4077–4088.

Asche, F., Misund, B., Sikveland, M., 2013. The relationship between spot and contract gas prices in Europe. *Energy Economics* 38: 212–217.

Boersma, T., Johnson, C., 2013. Twenty years of US experience: lessons learned for Europe. In C. Musialski *et al.* (eds), *Shale Gas in Europe: Opportunities, Risks, Challenges; A Multidisciplinary Analysis With A Focus On European Specificities*. Brussels: Claeys & Casteels Law Publishers.

Boersma, T., Khodabakhsh, C., 2014. EU engagement with shale gas. *Oil, Gas, and Energy Law Intelligence* 12(3): 1–12.

Boersma, T., Mitrova, T., Greving, G., Galkina, A., 2014. *Business As Usual: European Gas Market Functioning in Times of Turmoil and Increasing Import Dependence*. ESI policy brief 14-05. Washington, DC: The Brookings Institution.

Boussena, S., Locatelli, C., 2013. Energy institutional and organizational changes in EU and Russia: revisiting gas relations. *Energy Policy* 55: 180–189.

Christensen, T., Laegreid, P., 2007. Regulatory agencies: the challenges of balancing agency autonomy and political control. *Governance: An International Journal of Policy, Administration and Institutions* 20(3): 499–520.

CIEP, 2011. *CIEP Vision on the Gas Target Model*. The Hague: Clingendael International Energy Programme. See www.clingendaelenergy.com/inc/upload/files/Gas_Target_Model.pdf.

Coen, D., Thatcher, M., 2008. Network governance and multi-level delegation: European networks of regulatory agencies. *Journal of Public Policy* 28(1): 49–71.

Corbey, D., 1995. Dialectical functionalism: stagnation as a booster of European integration. *International Organization* 49(2): 253–284.

Correljé, A., De Jong, D., De Jong, J., 2009. *Crossing Borders in European Gas Networks: The Missing Links*. The Hague: Clingendael International Energy Programme.

De Hautecloque, A., Glachant, J.-M., 2009. Long-term energy supply contracts in

European competition policy: fuzzy not crazy. *Energy Policy* 37(12): 5399–5407.
Dorigoni, S., Graziano, C., Pontoni, F., 2010. Can LNG increase competitiveness in the natural gas market? *Energy Policy* 38(12): 7653–7664.
Elgie, R., 2006. Why do governments delegate authority to quasi-autonomous agencies? The case of independent administrative authorities in France. *Governance: An International Journal of Policy, Administration, and Institutions* 19(2): 207–227.
European Commission, 2010. *Communication: Energy Infrastructure Priorities for 2020 and Beyond: A Blueprint for an Integrated European Energy Network.* COM(2010) 677 final. Brussels: European Commission.
European Commission, 2012. *Communication: Making the Internal Energy Market Work.* COM(2012) 663 final. Brussels: European Commission.
European Commission, 2014a. *Communication on the Short Term Resilience of the European Gas System.* COM(2014) 654 final. Brussels: European Commission.
European Commission, 2014b. *Communication: European Energy Security Strategy.* COM(2014) 330 final. Brussels: European Commission.
GIIGNL, 2012. *The LNG Industry in 2012.* Paris: Groupe International des Importateurs de Gaz Naturel Liquéfié.
Johnson, C., Boersma, T., 2013. Energy (in)security in Poland, the case of shale gas. *Energy Policy*, 53, 389–399.
Joskow, L., 2005. Supply security in competitive electricity and natural gas markets. See http://economics.mit.edu/files/1183 (accessed September 11, 2012).
Kalicki, J. H., Goldwyn, D. L., (eds) 2013. *Energy and Security: Strategies for a World in Transition*, second edition. Washington, DC: Woodrow Wilson Centre Press.
Kumar, S., Kwon, H.-T., Choi, K.-H., Cho, J. H., Lim, W., Moon, I., 2011. Current status and future projections of LNG demand and supplies: a global perspective. *Energy Policy* 39(7): 4097–4104.
Larsen, A., Pedersen, L. H., Sørensen, E. M., Olsen, O. J., 2006. Independent regulatory authorities in European electricity markets. *Energy Policy* 34: 2858–2870.
Lochner, S., Bothe, D., 2009. The development of natural gas supply costs of Europe, the United States and Japan in a globalizing gas market: model-based analysis until 2030. *Energy Policy* 37(4): 1518–1528.
Maggetti, M., 2009. The role of independent regulatory agencies in policy-making: a comparative analysis. *Journal of European Public Policy* 16(3): 450–470.
Makholm, J. D., 2012. *The Political Economy of Pipelines: A Century of Comparative Institutional Development.* Chicago, IL: University of Chicago Press.
Paltsev, S., Jacoby, H. D., Reilly, J. M., Ejaz, Q. J., Morris, J., O'Sullivan, F., Rausch, S., Winchester, N., Kragha, O., 2011. The future of US natural gas production, use, and trade. *Energy Policy* 39(9): 5309–5321.
Pearson, I., Zeniewski, P., Gracceva, F., Zastera, P., McGlade, C., Sorrell, S., Speirs, J., Thonhauser, G., Alecu, C., Eriksson, A., Toft, P., Schuetz, M., 2012. *Unconventional Gas: Potential Energy Market Impacts in the European Union.* JRC Scientific and Policy Reports. Brussels: European Commission.
Renou-Maissant, P., 2012. Toward the integration of European natural gas markets: a time-varying approach. *Energy Policy* 51: 779–790.
Ruester, S., von Hirschhausen, C., Marcantonini, C., He, X., Egerer, J., Glachant, J.-M., 2012. *EU Involvement in Electricity and Natural Gas Transmission Grid Tarification.* Final report. Florence: RSCAS, European University Institute.

Spanjer, A. R., 2009. Regulatory intervention on the dynamic European gas market: neoclassical economics or transaction cost economics? *Energy Policy* 37(8): 3250–3258.

Stern, J., 2009. *Continental European Long-Term Gas Contracts: Is a Transition Away from Oil Product-Linked Pricing Inevitable and Imminent?* The Oxford Institute for Energy Studies, Oxford, NG 34.

Stern, J., 2014. International gas pricing in Europe and Asia: a crisis of fundamentals. *Energy Policy* 64: 43–48.

Szydlo, M., 2012. Independent discretion or democratic legitimization? The relations between national regulatory authorities and national parliaments under EU regulatory framework for network-bound sectors. *European Law Journal* 18(6): 793–820.

Thatcher, M., 2011. The creation of European regulatory agencies and its limits: a comparative analysis of European delegation. *Journal of European Public Policy* 18(6): 790–809.

Trombetta, J., 2012. *European Energy Security Discourses and the Development of a Common Energy Policy.* Working paper no 2. Groningen: Energy Delta Gas Research.

Vazquez, M., Hallack, M., Glachant, J.-M., 2012. *Building Gas Markets: US versus EU, Market versus Market Model.* Working paper RSCAS 2012/10. San Domenico di Fiesole: Robert Schuman Center for Advanced Studies, European University Institute.

Weber, J. G., 2012. The effects of a natural gas boom on employment and income in Colorado, Texas, and Wyoming. *Energy Economics* 34(5): 1580–1588.

Williamson, O. E., 2000. The new institutional economics: taking stock, looking ahead. *Journal of Economic Literature* 38: 595–613.

Index